D0004839

The Joy of Search

The Joy of Search

A Google Insider's Guide to Going Beyond the Basics

Daniel M. Russell

The MIT Press
Cambridge, Massachusetts
London, England

This book was set in ITC Stone Serif Std and ITC Stone Sans Std by Toppan Best-set Premedia Limited. Printed and bound in the United States of America.

Library of Congress Cataloging-in-Publication Data

Names: Russell, Daniel, 1955- author.
Title: The joy of search : a Google insider's guide to going beyond the basics / Daniel M. Russell.
Description: Cambridge, MA : The MIT Press, [2019] | Includes bibliographical references and index.
Identifiers: LCCN 2018059478 | ISBN 9780262042871 (hardcover : alk. paper)
Subjects: LCSH: Internet searching. | Internet research. | Web search engines.
Classification: LCC ZA4230 .R87 2019 | DDC 025.0425--dc23 LC record available at https://lccn.loc.gov/2018059478

10 9 8 7 6 5 4 3 2 1

This book is for my family:

for my mom, Marian, who showed me that asking questions is really, really important, especially when the answer might seem straightforward—and she taught me that it's often not;

for my kids, Chris and Katie, who keep me in touch with what's *really* happening on the internet, and for asking many questions that were tough to answer;

and for my wife, Lynne, whose "obvious" questions never seemed to have simple answers, but led me to entirely new avenues of investigation.

And to all of you readers: keep asking questions; they make life really, really interesting.

Contents

Acknowledgments

I'm grateful to Don Norman for saying, "You know, your SearchResearch blog really wants to be a book." He's the guy who really got this book project started with early guidance and encouragement. Thanks for the push.

Thanks also to Ira Chinoy, who also gave me a wagonload of encouragement in the process of figuring out what this book should be. We share a passion for research and archives—both valuable qualities in science and journalism.

Thanks to Sam Wineburg for pushing on the idea that we should understand how good searchers operate. His work and our conversations have been important in developing my understanding about how people do their research—and how they can improve.

Thanks to all the regular readers of my blog, SearchResearch. There are too many to name (and in many cases, I only know their blog names), but I want to mention frequent contributors Remmij, José Ramón Gonzalez, Jon-the-Unknown, Mike Ross, Rosemary, Anne Piascik, Debra Gottsleben, Passager, Luís Miguel Viterbo, Jeff Dowdy, Judith Koveleskie, Michael Michelmore, Clayton Lewis, and GRayR. There were many, many others over past four years; thanks to you all.

My appreciation to Mimi Ito for reminding me that most people think of online research as a pedestrian skill that shouldn't need any teaching. That's a humbling perspective that's completely right. She pointed out that this book needs to be intrinsically interesting. She's correct.

Thanks to the publication people at Google for letting me push the boundaries just a bit (with special appreciation to Peter Norvig, David Price, and Crystal Dahlen for reading everything!).

A huge thanks to my manager at Google, Scott Huffman, for allowing me to take the time needed to write this book.

I'm grateful to my family, especially my wife, Lynne, who always has the best ideas and critiques; my daughter, Katie, for copyediting help; and my son, Chris, for putting up with me during all the writing and endless questions.

In addition, I had a couple of good friends who invested a large amount of time in reading early drafts, and giving profuse and excellent commentary. Thanks to Robin Jeffries for valiantly reading early drafts of each chapter with depth and rigor. Luís Miguel Viterbo read with a valuable European perspective, which I deeply appreciated. Katie Russell has an eagle eye for copyediting—a skill that she did not inherit from me! Many thanks to Roz Foster, my literary agent, for walking me through my first book publication project (let's do some more!). Thanks also to Ann Awakuni Fernald, who has been wading through the texts with me, pruning, organizing, and tracking down permissions with great care.

And of course, a huge thanks to all the engineers, product managers, and researchers at Google. This book would not have been possible without your superhuman work.

1 Introduction: How You Can Harness the Power of Online Research—Why You Should Improve Your Online Researching Skills

This book has a simple idea: you too can be a great online researcher. It's not hard, and it's not computer geeky, but it's an incredibly useful skill to have. If you can do better online research, you can ask and answer questions better—you can be more accurate, and you find out the information you need faster and more completely.

I'm a professional researcher who works at Google on the team that built the search engine. A big part of my job is to understand how people are using Google, in both good ways and less good ways. How do they use it successfully? And how is it that things occasionally don't work the way you'd expect? For instance, sometimes I'll see searchers spending thirty minutes searching for something that should take less than two minutes. Naturally, sometimes they find a less than great result. What makes someone a great online searcher? This book makes the case that learning how to be a better online researcher isn't just a good idea; it's something that we all need to know how to do.

Consider this: we all know about the three R's of education: reading, writing, and 'rithmetic—the three basic skills that schools have to teach (which obviously doesn't include spelling). This book makes the case that there's a fourth R that we should be teaching to all students and adults: that is, *research*.

I don't mean that you'll have to spend your time doing learned investigations into strange and inscrutable depths of obscure history, or weird and exotic scientific studies. What I DO mean is that everyone can learn to use online information much more effectively and efficiently, whether you're a student, citizen, parent, or professional researcher; there are things you should know about online research.

If you think about it, learning has changed from a school-only activity to a lifelong activity. And just as the person who can learn the best and know the most has a real advantage, so too does being a great online researcher give you a real advantage—in school, your work, and life. People fluent in search not only save time but are also far more likely to find higher-quality, more credible, more useful content. More important, they can ask questions that were impossible just a few years ago.

As Samuel Johnson wrote, *Knowledge is of two kinds. We know a subject ourselves, or we know where we can find information upon it.*[1]

While that's true, the common version of his quote usually leaves off the rest of his paragraph: *When we enquire into any subject, the first thing we have to do is to know what books have treated of it. This leads us to look at catalogues, and at the backs of books in libraries.*[2]

In Johnson's day that meant knowing how to do research AND knowing that catalogs existed, that libraries were collections of books on topics of interest, and that the back of a book contains an index. It also meant that you knew **how** to get into a library, many of which were still private and by subscription (that is, by "invitation") only.

Most of the searches that Google sees in a typical day are fairly straightforward. The goal is clear, and the results are pretty obvious and unambiguous.

But a significant number of searches are not. Searchers might have a goal in mind, but they can't figure out how to express it in a way that will give them what they want. Sometimes their query is precise, but they don't know how to read and interpret the results. It drives me to distraction as a researcher because I know that the searcher is missing just one small but critical piece of information. We try to build as much as we can into the search algorithm, but people still need to know a bit about what the web is, and how search engines crawl, index, and respond to their queries. (Unlike in Johnson's day, there's no index in the back of the book of the internet.)

In a sense, that's my mission: to help people become better researchers, beyond just the basic skill of knowing how to make Google dance to their tune. My goal is to help people understand the larger issues at play here: what it means to be a literate person now, and how to continuously learn to be literate as changes happen in the future.

Research is a skill that we all take for granted—we all use Google, and so we assume we're good at it. Yet online research is a critical skill for our future. As the nature of work and education changes (and that, really, is the only constant we have), we need to bring our students (and ourselves) up to speed on what it takes to be good searchers. As teachers, parents, and citizens, we need to give them the skills of the fourth R—**research**—and all the skills and knowledge they need to function effectively as learned searchers.

What's more, we're trying to equip them with skills they can use not just now but also for every information search problem that we confront now and in the future.

So what kinds of things do you need to know?

To start with, there's so much information available online that you might be forgiven for thinking it's all online somewhere. And it is truly amazing what *is* available with a quick Google search. We live in a time of fantastic information richness and depth. But it wasn't always like this. And what's more, you need to know where the limits are, and how to detect when you've hit them.

One of the great discoveries of my young life was when I learned that there are books with answers in them that could satisfy my curiosity and let me do research almost instantaneously. Since I was a curious child, I found that by just picking up a volume of an encyclopedia.[3] I could read about many wonderful things—Eritrea, Eisenhower, eggs (development of), Earth (planet), ecru (the color) ... on and on. All these riches—and just in the volume of the letter "E." What mysteries lay in "F" or "G"? Why was there a volume "X, Y, and Z," and not "M, N, and O"?

That was then, this is now, and it's difficult to find any reference content that's organized alphabetically. Why would you, when it's so much faster to do a search for what you seek? Today we have vast resources online, indexed by search engines, or sometimes organized into websites and databases—all there for the questing mind to use.

All that information is out there, so it's frustrating to see books and articles that assert some piece of information is true, but don't tell you the source or how they know it's true.

I find it deeply frustrating to be told something about the world and then have it be unsubstantiated. I've always wanted to know how I could figure this out on my own.

This is especially true for books of trivia and fascinating content. You know what I mean—those books and websites that collect the biggest, longest, heaviest, or most unbelievable facts about the world. They're great, but I want to shout at them, HOW do you know that? WHERE did you find that morsel of information? (Or did you just make it up?) Even fact-checking websites and collections of interesting facts don't tell you the story behind the information. If you're lucky, they'll point to something else. But it's rare that they actually tell you how to figure it out for yourself. And that's only if you're lucky.

There are also a good number of books that tell you how to use online information resources. But there's a gap in those texts: they tell you the various things you can do, but not how to hook them all together to go from an initial question to a satisfying answer. This book, however, is about HOW to answer those questions yourself. It is about HOW to frame your questions and use the various online research tools that let you get the answers yourself, from the beginning of your curious question, to finding results via many different online resources.

And that's the point of this book: you too can ask, and answer, your curious questions.

Here I'll show you how to find things out by showing you the steps (and occasional missteps) that I've gone through while doing my own research. Each chapter is a story—one that documents how I've found the answers to questions I've had (or have been asked). Along the way, I'll tell you exactly *how* I did each search step, and *why*.

My work for the past thirteen years as a research scientist at Google has been to study how people use various kinds of online information sources to answer questions that they have. And it's been puzzling. The online world has an immense set of resources at hand, and yet I see that most people don't really understand how to use it effectively. It's as though we've all been given a Steinway, but we only know "Chopsticks." Or you just got a Formula 1 racing car, but nobody ever showed you how to take it out of first gear. We have these amazing tools that we just don't know how to use well.

For most simple queries—those in first gear—search engines (such as Google) work extremely well. Search engines' ability to quickly provide answers to these queries is a remarkable testament to their power and scope as well as the advances in information engineering during the past few decades. But there's so much more these engines can do.

Clearly, search engines are different from traditional information sources; they're not at all like the encyclopedias of my youth. Need to know the population of Japan? The number of elementary schools in the United States? The signers of the Declaration of Independence and where they lived at the time? How about the distance from the Earth to the Sun? Ask, and you'll get the answer in the twinkle of an eye. With the rise of easily accessible online searches, anyone can answer these questions quickly, accurately, and at any time of day from their laptop, desktop, or phone. What's more, unlike the encyclopedias of yore, these information sources are up to date.

And yet if you know just a little more about searching, if you realize the depth of information processing available to you, if you learn a touch of sleuthing and take advantage of your new ability to shift ordinary memory tasks to the machine, search engines can become more of an intelligence amplifier for you than just a simple collection of data. That power is not only truly *fun* to have, but also gives you the ability to understand how things work and see more deeply in the world.

Through online search tools, the world is your information oyster, at your disposal twenty-four hours a day, seven days a week. Those who are fluent in search and retrieval not only save time but also are far more likely to find higher-quality, more credible, more useful content. They can ask questions that were impossible just a few years ago. They can think faster and find solutions more easily. They can quickly get to the bottom of that wild news story they read and the top of their career ladders too. Whether they're students, professionals, or stay-at-home parents, those who are good at doing online research have a real advantage over those who aren't.

Even though "research" is often thought to be what professional scientists do, the truth is that it's something we all do every day. My mission—the mission of this book—is to help you become better at it, go beyond the basic skill of typing something into Google and learn how to ask a search engine a good question, and understand what it means to be a literate person today and in the future. My goal is to show you the marvelous depth of knowledge at your fingertips, and my hope is to share with you the great joy of finding out.

I'll also talk about the attitudes and research styles that will make you more productive and accurate. For instance, a common misconception is that you can search and learn about anything, starting with nothing. The reality is that many research quests are foiled by not knowing enough about *how* to search. There's a bit of a skill here that we'll talk about. I'll talk about how to start learning something in an area about which you know almost nothing. Unsurprisingly, to learn something, you have to begin with something. But how do you do that when your knowledge is at zero? *Chance favors the prepared mind.* Why is that true? And how can you prepare your mind?

There are skills and attitudes to learn from this book. My goal is to teach both: the skills to let you be an effective researcher, and a person with research-y attitudes. Both are important.

Mostly, I HOPE to make you curious about the world and help you learn some of the skills it takes to successfully do online searches to get to good, high-quality, believable answers.

How to Read This Book

This book is composed of sixteen stories about how I answered some questions that came up for me during the course of my life. These chapters walk you through my stories of discovery, laying out the journey I went through to find out the answers to some particularly interesting questions. Along the way, I'll show you just how I solved them, and how you can do the same kind of thing.

The first half (chapters 2–9) are primarily about tactics—that is, the individual steps you need to know to work with online resources. In these chapters, I'll use a variety of different resources along with some of the tips and tricks that make searching these easier.

In the second half (chapter 10–18), I'll talk more about the strategies I used to guide my search process. When should you keep searching? When should you stop? How do you decide when a particular research path is leading you to a dry hole?

In each chapter, I'll start our investigations with a **research question**, using this as a way to be careful about what we're searching for. You'll see each research question separated out in the text. For example, "Where was this photograph taken?" Or the research question might be some curiosity about some observation in the world—such as "Why do lakes sometimes explode?"—that seems utterly bizarre, and when you follow your curiosity, you'll find out where, why, or how this thing can be answered.

Even though it's a simple thing, just writing down the question has the side benefit of helping to *frame* what you're about to do. I know, it sounds too simple to actually work, but it's true. If you take ten seconds to write down your question BEFORE you start your research process, you'll find your process will be much, much more effective. It's almost magical; you have to figure out (a) what you're really asking, (b) what terms you'll be searching for, and (c) what kind of answer will satisfy you.

That's what I do in each chapter: I write down the questions, and then walk through the steps I took to find the answer. Instead of just *telling* you the answer, I'm trying to make it clear exactly how I went from *really interesting question* to an answer that I believe.

As you read these stories, I'll point out how I use each of the tools and information resources along the way. Every so often you'll see some text in square brackets like this: [exploding lakes in Africa]. That's a query. The text in the brackets is what I searched for in order to answer a particular question. But note that you shouldn't include the brackets in your search; they're just there to show you where the actual query starts and ends. Every chapter has the online searches I used along the way.

At the end of each chapter is a set of research lessons that summarize the things you should have picked up during the story. These are the insights that you can use in answering your own questions.

I've also added a bunch of "How to Do It" notes at the end of each chapter. When you see bolded text that has a superscript letter—such as this example of how to make a good query for **searching**[A]—you'll find a short note at the end of the chapter telling you how to do this particular step of the research process. (And of course, if that's not enough information for you, you could always do another search to get even more details!)

As you read you'll also see URLs that look like this: bit.LY/TJOS-1-1. These are links to online web pages that are easier to type than a full URL, which are often long and full of strange characters that nobody can transcribe correctly. But the **bit.LY** links are short and easy to type. This is a nice solution to the perennial problem of "How do you put links to online resources in a print document?" Note that these links are case sensitive. TJOS-1–1 is not the same as tjos-1–1! Be sure to use all capital letters.

Note also that you should not include any trailing periods. In the typesetting process, a period is added to the end of a sentence to indicate the end. But if that sentence (or citation) ends with a URL, be careful not to add the trailing period to the URL you type! You want to type bit.LY/TJOS-1-1 without any extra period (i.e., do not type bit.LY/TJOS-1-1.).

The last two chapters summarize what this book has to teach. It's really true that in this case, the whole is greater than the sum of the parts. Chapter 19 offers many of the lessons I've learned all pulled together into a tidy summary that's worth reading because it reflects on ALL the chapters. Sometimes you can see things more clearly when you see many of the lessons side by side. Chapter 20 gives you an idea about the future of online research.

Now let's go find out something amazing about the world.

I'll start the next chapter with a photograph that a friend sent to me, asking where on earth he took it. Remarkably, I was able to figure it out fairly quickly just by doing a few online searches. Let's see how. Once you see how it's done, you'll be able to do this too.

Onward. Let's do some research.

How to Do It

A. How to make an effective query. One of the great pieces of search engine technology is that queries can be simple, just a few words long, and you'll often find the information you're seeking.

Knowing a couple of things about how search engines work, however, will make your searching far better.

You should know that there are a few different approaches to doing a query on a search engine such as Google, Bing, Wolfram Alpha, or DuckDuckGo.

The *classic* query style is to use a few words (sometimes called *keywords*) that describe the kind of web result you seek. A classic query is best made by choosing a few terms that are central to the idea you're searching for, and ones that you think will be on the web page. Classic queries look like this:

[Japan population]
[most Olympic gold medals]

As search engines improve their question-handling ability, they are getting better at handling questions posed in a more natural fashion, such as a *question-answering* query style like this:

[What is the population of Japan?]

[Who has won the most Olympic gold medals?]

Generally speaking, the best queries are those that use on-topic words (for the classical queries) or consist of clearly stated questions. When a question-query is answerable, the results usually give much more context information on that topic. (For instance, [What is the population of Japan?] will give some contextualized data as well.)

While search engines let you use some special operators (that is, special terms that modify the meaning of your query), more than 99 percent of all queries don't use anything special; they're just classical queries or simple questions.

In particular, people rarely, if ever, use *Boolean queries* in their classical searches.[4] Both Google and Bing automatically put an implied AND between all the terms (although they might drop a term from the list if there aren't any results with all the terms). On the other hand, sometimes using an OR in the query is a useful way to specify synonyms for a search term. For example, you can look for jobs in either (or both) of two cities with a query like this:

[Dallas OR Houston jobs software development]

This query means that the search engine will look for either Dallas *or* Houston on the web pages.

But if you don't want anything from Houston, you can use the minus sign (–) before that term to exclude it from the search results. This query will show jobs in Dallas, but none from Houston.

[Dallas –Houston jobs software development]

Overall, the classical queries are best when they use terms that you'd expect to find on the ideal page that will answer your query. There's much more to talk about, and that's what the rest of the book will discuss.

2 Finding a Mysterious Location Somewhere in the World: How to Use Multiple Information Sources to Zero In on a Resource

Where in the world is this? How can you figure it out? A few tips to get started on doing really surprising searches.

If you work at Google, you end up attracting all kinds of questions. Some are just requests for help, often in a panic ("Can you find this kind of data for me ... by 5:00 p.m.?"). But sometimes you get little puzzles ("Bet you can't find ...") or pictures of someplace in the world with the question, "Where's this?"

Here's a picture that a friend sent me, asking not just where in the world this office tower is, but also an apparently difficult question to answer just from the photo; that's the question for this chapter.

Research Question: *Where was the photographer standing when he took this photo?*

This seems impossibly hard, but with a little ingenuity, it's quite possible. In under five minutes, I was able to go from just this photo to his location (and for extra credit, I was able to find the phone number of the office where he was standing).

There are several ways to do this, but the simplest one is to start with clues you find in the photo itself. First, ask yourself, What do you see in the photo?

In a case like this, you have to start with what you have. Look at the image carefully, noting what you might use to crack this where-in-the-world challenge.

Figure 2.1
A mystery picture sent by a friend, who asks, "Where was the photograph taken from?"
Credit: Daniel M. Russell

There are only a few things that you can really begin with in this picture. Notice that two buildings have logos that might be useful, and there's a flag in the lower-left corner. If you look *very* closely, you'll see it's actually a mural of a postage stamp with a cancellation mark (figure 2.2). Here's what that looks like when you **zoom in**[A]. (As mentioned earlier, when you see bolded text that has a superscript, it's telling you that there's a short summary of how to do this at the chapter's end. Check it out; at the end of this chapter is a how-to about ways to zoom in on an image.)

Otherwise, it's a fairly ordinary urban scene.

A quick check of the flags of the world (just do a quick search for [flags of the world] in Google Images, and you can quickly get a lovely diagram with all the world flags). (See figure 2.3.)

You can click on any of these and without much trouble find that this is an oddly cropped Egyptian flag. Is this somewhere in Egypt? A quick visual inspection of the buildings and lay of the city tells me this surely isn't Cairo

Figure 2.2
Zoom in on postage stamp in Warsaw.
Credit: Daniel M. Russell

or anywhere else in Egypt, so that's just a false lead.[1] So let's ignore the flag for the moment and go looking around in the image for more clues.

Let's try searching for one of the two logos. I searched first for the closest and most obvious logo atop the building labeled "TP." (I ALSO chose this because the building is clearly pretty new and pretty much a trophy building. I figured there would be a good chance that someone would have put up a web page on it.) I tried queries like [tp logo], but didn't have much luck until I modified the query just a bit with a *description* of what I was looking for.

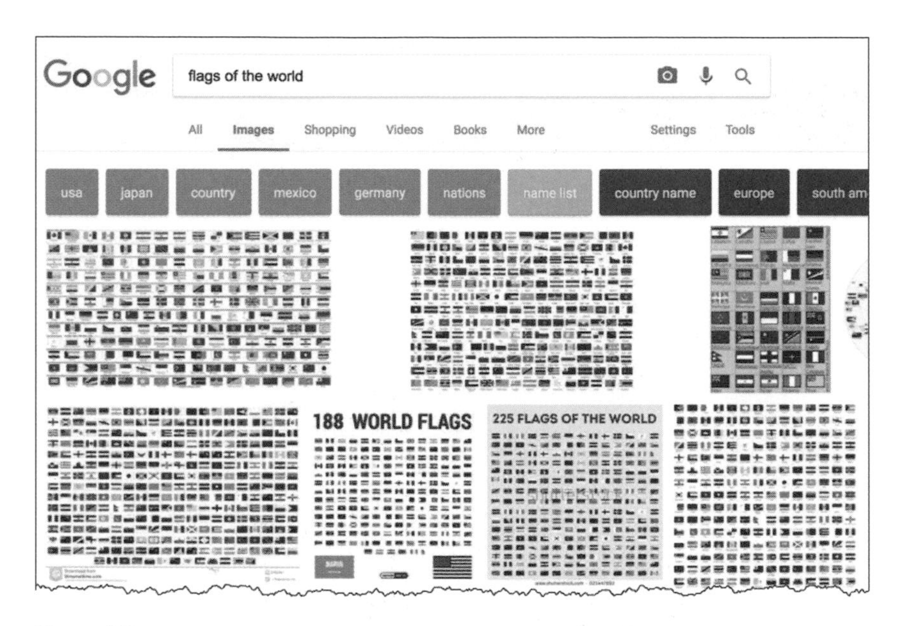

Figure 2.3
A search for [flags of the world] in Google Images gives many grids of flags from around the world.
Credit: Google and the Google logo are registered trademarks of Google Inc., used with permission

I searched for [tp office building] and voilà! This is pretty clearly the Telekomunikacja Polska building in Warsaw, Poland. The first result on the search page was to a site that had all the skyscrapers of Warsaw labeled, and this is clearly that building. A quick Wikipedia search for confirmation, and you'll see all is lining up; this is the TP building, which is now called the "Spektrum Tower (formerly TP S.A. Tower)." (See figure 2.4.)

You can also figure this out by doing a Google Images search for the name of the company [Telekomunikacja Polska]. The TP logo shows up on the first page of Google Images results, and that leads you to the Telekomunikacja Polska website.

Knowing this, I could easily get the street address for the TP building: 14/16 Twarda Street, Warsaw (figure 2.5).

Of course, there are always multiple approaches to finding the location of a photo, and cityscapes are full of clues.

If you look carefully at the building behind the TP skyscraper, you'll see another building with another logo—PZU. A quick search for [PZU] or

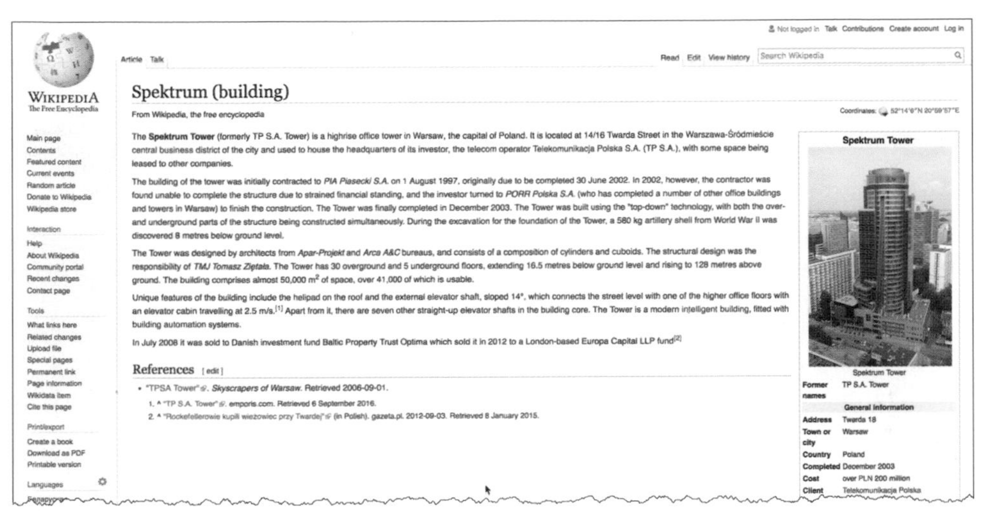

Not logged in Talk Contributions Create account Log in

Article Talk Read Edit View history Search Wikipedia

WIKIPEDIA
The Free Encyclopedia

Main page
Contents
Featured content
Current events
Random article
Donate to Wikipedia
Wikipedia store

Interaction
Help
About Wikipedia
Community portal
Recent changes
Contact page

Tools
What links here
Related changes
Upload file
Special pages
Permanent link
Page information
Wikidata item
Cite this page

Print/export
Create a book
Download as PDF
Printable version

Languages

Spektrum (building)

From Wikipedia, the free encyclopedia

Coordinates: 52°14′6″N 20°59′57″E

The **Spektrum Tower** (formerly TP S.A. Tower) is a highrise office tower in Warsaw, the capital of Poland. It is located at 14/16 Twarda Street in the Warszawa-Śródmieście central business district of the city and used to house the headquarters of its investor, the telecom operator Telekomunikacja Polska S.A. (TP S.A.), with some space being leased to other companies.

The building of the tower was initially contracted to *PIA Piasecki S.A.* on 1 August 1997, originally due to be completed 30 June 2002. In 2002, however, the contractor was found unable to complete the structure due to strained financial standing, and the investor turned to *PORR Polska S.A.* (who has completed a number of other office buildings and towers in Warsaw) to finish the construction. The Tower was finally completed in December 2003. The Tower was built using the "top-down" technology, with both the over- and underground parts of the structure being constructed simultaneously. During the excavation for the foundation of the Tower, a 580 kg artillery shell from World War II was discovered 8 metres below ground level.

The Tower was designed by architects from *Apar-Projekt* and *Arca A&C* bureaus, and consists of a composition of cylinders and cuboids. The structural design was the responsibility of *TMJ Tomasz Ziętala*. The Tower has 30 overground and 5 underground floors, extending 16.5 metres below ground level and rising to 128 metres above ground. The building comprises almost 50,000 m² of space, over 41,000 of which is usable.

Unique features of the building include the helipad on the roof and the external elevator shaft, sloped 14°, which connects the street level with one of the higher office floors with an elevator cabin travelling at 2.5 m/s.[1] Apart from it, there are seven other straight-up elevator shafts in the building core. The Tower is a modern intelligent building, fitted with building automation systems.

In July 2008 it was sold to Danish investment fund Baltic Property Trust Optima which sold it in 2012 to a London-based Europa Capital LLP fund[2]

References [edit]

- "TPSA Tower". *Skyscrapers of Warsaw*. Retrieved 2006-09-01.

1. ^ "TP S.A. Tower". emporis.com. Retrieved 6 September 2016.
2. ^ "Rockefellerowie kupili wieżowiec przy Twardej" (in Polish). gazeta.pl. 2012-09-03. Retrieved 8 January 2015.

Spektrum Tower

Spektrum Tower

Former names	TP S.A. Tower
General information	
Address	Twarda 18
Town or city	Warsaw
Country	Poland
Completed	December 2003
Cost	over PLN 200 million
Client	Telekomunikacja Polska

Figure 2.4

A Wikipedia search for [TP] leads to the Spektrum Tower in Warsaw.

Credit: Cezary Piwowarski, Wikipedia, used under a CC BY 3.0 license

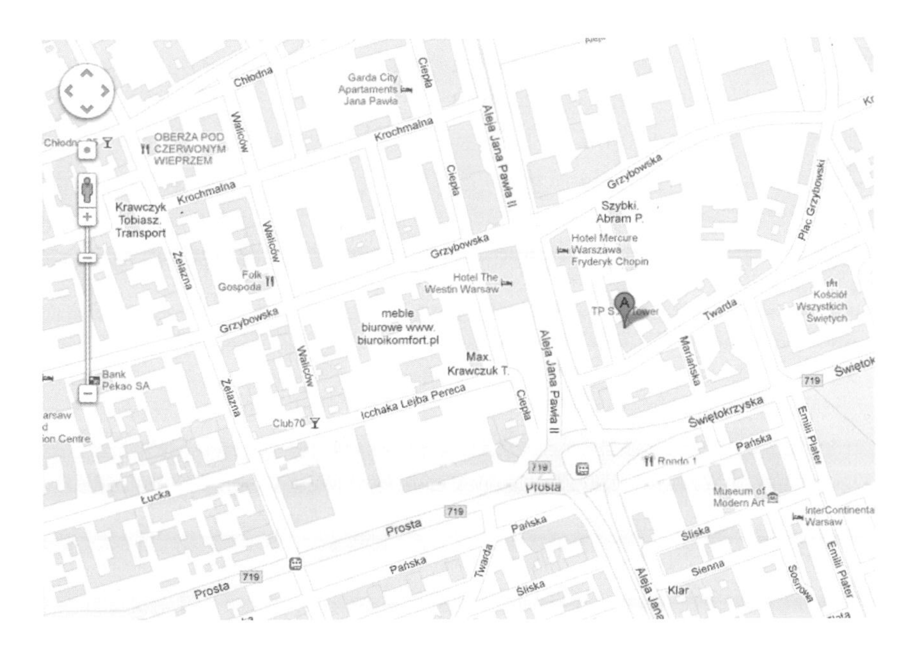

Figure 2.5

The red pin shows us where 14/16 Twarda Street is in Warsaw, but now where's the building we seek?

Credit: Map Data © Google

[PZU building] also leads directly to Warsaw, and the Powszechny Zakład Ubezpieczeń (PZU) tower, just a few blocks away from the TP tower.

Now that we've narrowed the location down to Warsaw, how can I find where the photo was taken FROM?

I tried doing this with Google Maps and looking around downtown Warsaw with **Google Street View**[B], but after a couple minutes I couldn't quite get the angle I wanted to verify where the photo was taken from. That's when I realized this was really a problem perfectly suited for **Google Earth**[C].[2]

Searching in Google Earth for the TP building at 14/16 Twarda Street, I was able to get to exactly the same view I see in Google Maps. But if you turn on the 3D buildings layer, you get a different perspective. Not only can you see all the satellite imagery, but you can see all the major buildings in the downtown area.

I then just used the camera/viewer controls and flew around in Google Earth until I pretty much matched the view. Finding the right side of the building was simple, and then it was just a bit of moving the viewpoint around until I pretty much aligned the Google Earth view with the image.

Once I had the view matched with the photograph (which I did by lining up the buildings), I literally "turned around" my view in Google Earth to see the building behind me—that is, the one where the photograph was taken (figure 2.6).

In Google Earth, double clicking on a 3D building will give you information about that building. I double clicked on the building, and up popped the information I was looking for; it's the Warsaw Financial Center.

Now ... how to figure out what office/floor you're on? You can estimate it by looking at the altitude of the view—something that Google Earth tells you in the lower-right corner. From that number (120 meters), we can guess we're somewhere around the fifteenth floor or so. But to get the exact floor? Ah ... THIS takes a little extra detective work! How can I figure that out?

Easy, but you have to pay attention to some of the details.

When I opened the large image, the first thing I did was to scrub around a bit, looking for what I might use for clues. In the process, I noticed an odd blue set of squiggles on the image roughly in the middle of the photograph. That's a little puzzling. What's up with that? I ignored it earlier, but now I remembered it was there and so I zoomed in a bit to see (figure 2.7).

Figure 2.6
Flying around in Google Earth, I was able to find the location in Warsaw where the Google Earth view with the 3D buildings view mode looks nearly identical to the original image.
© 2012 CNES, Daniel M. Russell

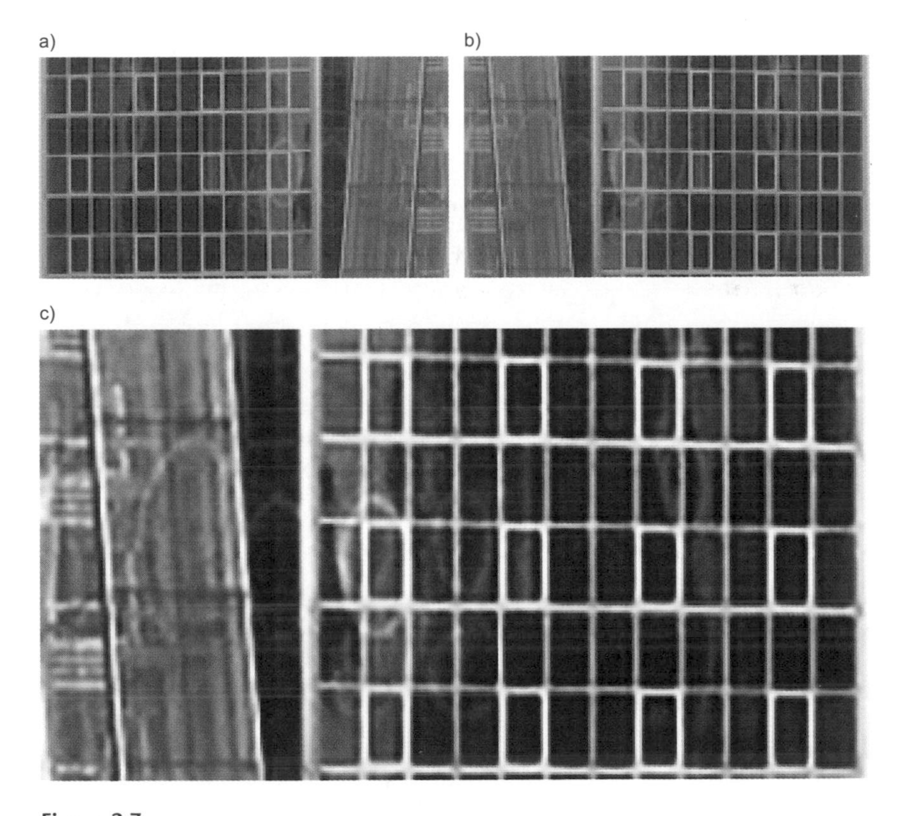

Figure 2.7
a) The original photo cropped down to just the blobby reflection in the window. b) If you flip that image horizontally (as though looking at it in a mirror), a word appears. c) Zoom in a bit and then enhance the contrast, and it's clear what the logo is.
Credit: Daniel M. Russell

This looks suspiciously like backward writing. At this point, I needed to play around with the image a bit to see if I couldn't squeeze some more information out of it. So I opened this subimage in my favorite image editor and **flipped it around the vertical axis**[D] (in this case, I used Adobe Photoshop, but nearly all image editors can do this).

Since it looked backward, I first did a flip horizontally so I can see what it could be (figure 2.7b). At this point, not only does it look suspiciously like writing, but it looks suspiciously familiar. Still, I wasn't 100 percent sure. I needed a better look.

Just to be sure, using my image editor I enhanced the contrast a bit and zoomed in so I could see the details (figure 2.7c). This is definitely the Google logo.

And then I had my AHA! It's the Google logo! (Surprise!) Obviously, my friend was in the Google office in Warsaw when he took the photo. Offices often have large logos of their company, so this isn't that surprising. (But now I get the humor: my friend doesn't work at Google, and so this was a bit of a "look where I'm at now!" kind of inside joke.)

From there, it's not hard to do a search for [Google office Warsaw] and find the Google offices in the Warsaw Financial Center on the Google offices web page. That page tells us the exact floor (tenth) and phone number.

Search [Google offices Warsaw] to find that the information is at the Google corporate offices website: www.google.com/about/locations/.

Google Warsaw

Warsaw Financial Center

Emilii Plater 53

00–113 Warszawa

Poland

Phone: +48 22 207 19 00

Fax: +48 22 207 19 21

And once you have THAT information, it's pretty straightforward to figure out that the Google offices are on the tenth floor.

Research Lessons

1. *Sometimes clues can be misleading (such as the Egyptian flag in the image).* It's important to NOT get bogged down in dead ends but instead be willing to change your strategy on the fly.

2. *Google Earth is a valuable search tool when you're looking for objects in landscape photos that you can't identify otherwise.* In this case, I was able to line up the photo with the 3D buildings and literally work backward to identify the Warsaw Financial Center.

3. *Sometimes clues are hidden in the details and reflections.* In the case of the Google logo appearing reflected in the glass, it becomes apparent once you isolate that part of the image, magnify it, and reflect it around the horizontal axis. That one clue then lets you figure out the exact location. Reflections are often incredibly useful clues in images (although you need to have an image editing tool to magnify the details, transform the image, or enhance the contrast).

Above all, you just have to think like a detective, working from the clues you have (logos and the reflection on the glass) and then using the tools you know about (such as Google Earth) to work from evidence to the complete answer.

How to Do It

A. Zoom in. To *zoom in* on a photo, you can zoom in on an image by using the browser itself. Just type a control-+ (that is, hold down the control key and press the + symbol) to zoom in and then control-– (the control and minus keys). On a Mac, it's similar, but use the command key instead of the control key (that is, CMD-+ and CMD-–).

If you need even more zooming, you can download the photo from the website and then use an application to magnify the image. For most browsers, you can right-click on the image (or control click on a Macintosh) and then save the image to your computer. Then once you've done that, just double click on the image, and your computer will open it in an image editing program like Apple's Preview or Microsoft's Photo Viewer. No matter what your image viewing application is, there will be an option to zoom in, scroll around, and zoom out. Often the zoom-in tool looks like a small magnifying glass, frequently with a plus sign in it. Click on that to zoom in. (And if it's not there, it will be hidden under the menu item "Tools." Look there as well.)

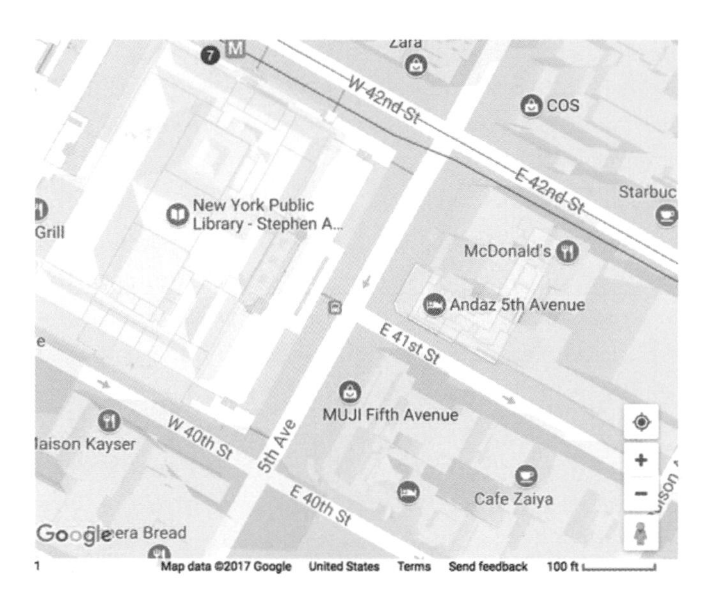

Figure 2.8

B. Google Street View. Using Google Street View is the way you can "look around" on a street location when you use Google Maps. To do this, just pin Google Maps on your browser in whatever location you'd like to check out. Here I've opened it up to Manhattan, New York (figure 2.8).

You use Google Street View by clicking and holding the small yellow man figure in the lower-right corner. Once you do this, you'll see that you can drag that icon onto any of the blue lines that appear on the map (figure 2.9).

If I drag the icon to the corner of East Forty-First Street and Fifth Avenue, and then drop the icon at that map location, I'll be able to see the front of the New York Public Library (figure 2.10).

You can then click and press on the image to change the view by dragging the view left, right, up, and down. (Try it!) Much of the world's streets are visible this way. (Check out your home street, or try someplace you might want to visit but can't get to at the moment. 1600 Pennsylvania Avenue in Washington, DC, is popular, but you can also check out the Google Street View of a place like the Eiffel Tower in Paris, France (figure 2.11).

C. Google Earth. Google Earth is a free application that you can download from Google. (I could give you the URL, but a better way to find it is for you

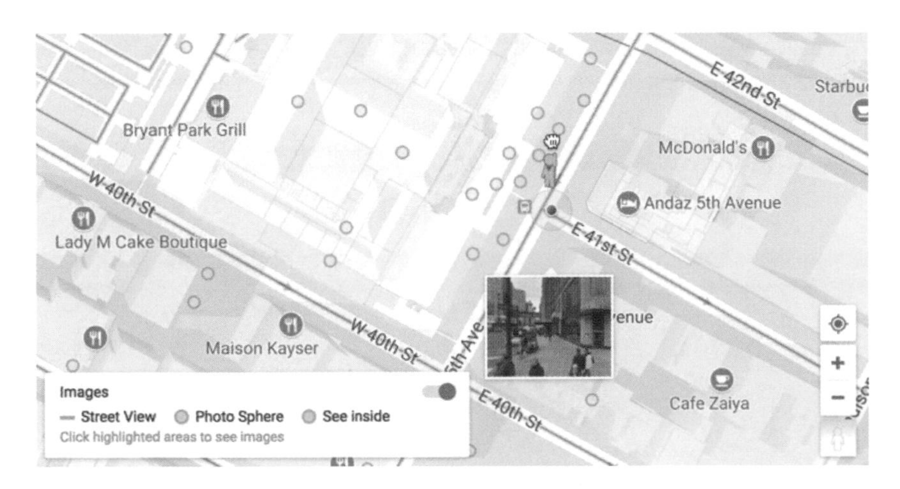

Figure 2.9
© 2017 Google

Figure 2.10
The Google Street View of the New York Public Library.
© 2014 Google

Figure 2.11
Using Google Street View, you can visit Paris and wander the streets without having to buy an airline ticket.

to search for [Google Earth] and then download it from there.) With Google Earth, you can literally fly to any spot on the planet and then look around (figure 2.12). It's much like Google Maps, but it has a bunch of extra features that Google Maps doesn't have (at least not yet). For instance, you can turn on (or off) various "layers" of mapping information—city and country outlines, or the locations of different kinds of landmarks. In this chapter, I've already talked about the 3D buildings layer, which is an information layer that you can turn on or off when looking at the Google Earth images.

In figure 2.13, we see the River Thames, Big Ben, the Houses of Parliament, and Westminster Abbey just behind. This is the same kind of view I used to locate that building in downtown Warsaw.

D. Flip it around the vertical axis. Knowing how to manipulate an image is a generally useful thing. Not only does it let you alter the image to your liking, but as a side effect, you learn what *other* people can do to change an image. As you probably know, image manipulation is rampant in the fashion world, and happens more often than you might like in news and journalism.

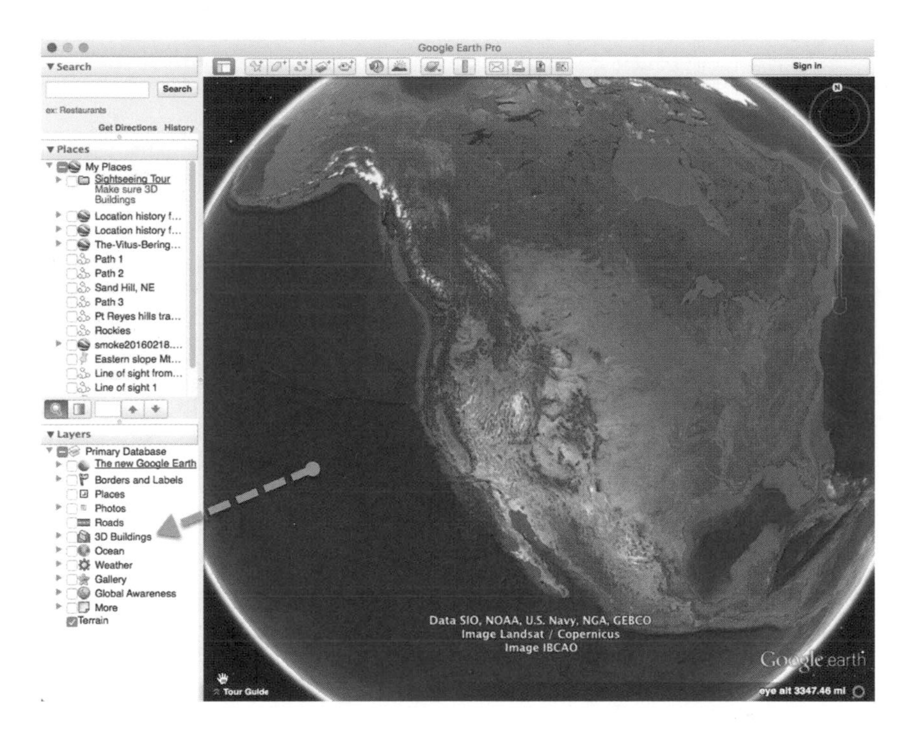

Figure 2.12
In Google Earth, you can click on the small box and activate the 3D buildings view. That layer lets you see buildings in 3D from around the world.
Image © Google. Map data: Landsat/Copernicus/IBCAO, SIO, NOAA, US Navy

If you open a picture in almost ANY image editor, you'll see a tool (frequently hidden in a pull-down menu called "Tools") that lets you flip the image.

To flip an image, you have to know that there are two different ways to flip—*horizontally* and *vertically*. Here's an image of a young woman looking to the right (figure 2.14).

And after doing a "flip horizontal," the image looks like the same picture but everything is flipped left to right. She now looks to the left.

If you do a "flip vertical," she looks to the right.

Note that rotation is NOT the same as flipping the image. If you rotate the image by 30 degrees, you'll see something different.

And notice that rotating the original image by 180 degrees gives you a DIFFERENT image than if you do a flip vertically.

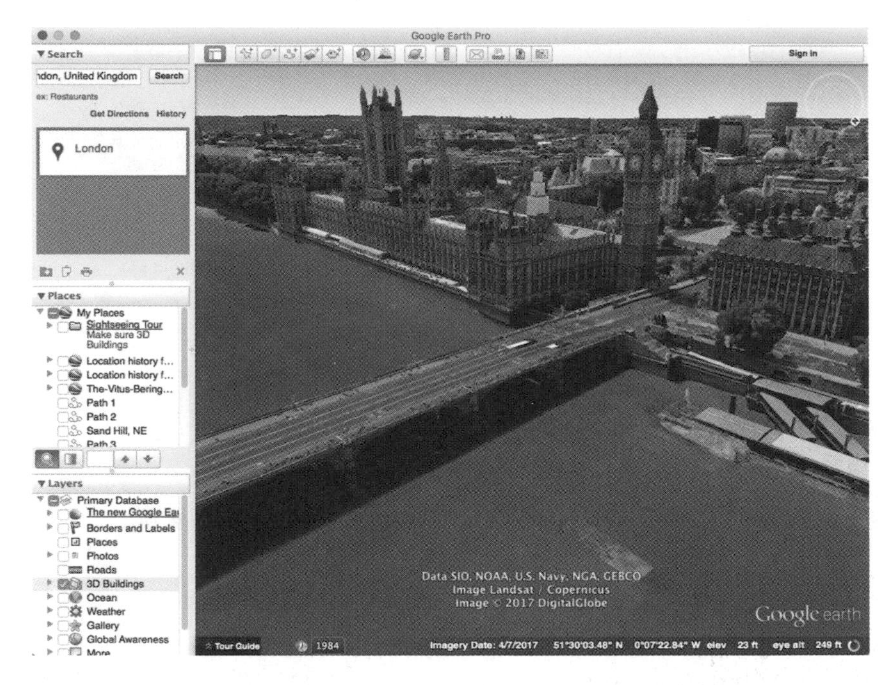

Figure 2.13

A view in Google Earth of the River Thames, Big Ben, the Houses of Parliament, and Westminster Abbey. To get here, search for [Parliament London UK] with the 3D buildings layer activated.

Image © Google. Map data: Landsat/Copernicus/IBCAO

In the case of the photo featured earlier in this chapter, I wanted to make it possible for me to recognize the image that was faintly visible in the window. To "undo" the effect of the reflection, I had to do a flip horizontally in order for it to show up in the right way.

In this particular case, I used the "preview" application on the Macintosh (but again, nearly all photo viewers have this capability).

To really understand this, I highly recommend that you open an image on your computer and play around with the flip function (and the rotate function, if you have a couple of minutes) to really understand what's going on here. Trust me, this will be useful for you in the future.

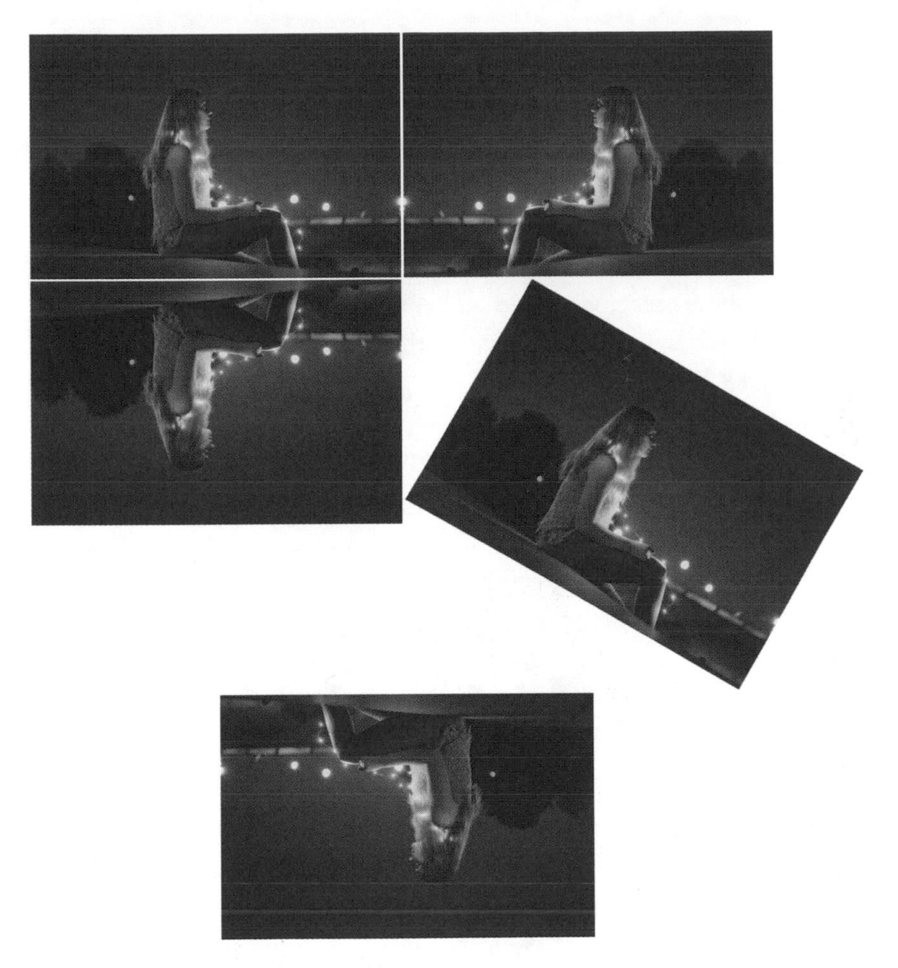

Figure 2.14
The upper left shows the original picture, before any rotating or flipping, and then various orientations.
Credit: Luc Nikiema

Try This Yourself

Here's a lovely picture I took of the Notre-Dame Cathedral in Paris, France. As you can see, it's a beautiful spring day with clouds and birds, and a bridge crossing the river. Imagine that you have an incredibly romantic episode in your life that happened on that bridge and that now—years later—you'd like to do a little research about the bridge. You know the name of the cathedral in the background, and probably know the name of the river, but how would you figure out the name of the bridge? (Hint: use Google Street View.) Can you search it out?

Figure 2.15
Credit: Daniel M. Russell

3 Do Lakes in Africa Sometimes Explode? How to Focus Your Search with "site:" and Using Specialized Terms

Can some lakes erupt and kill thousands of people? How would you find out?

Not long ago I was reading about a strange lake. I know what you're thinking: How strange can a lake be? A lake's a lake, right? In this case, the lake I'm thinking of seems not just strange but also deadly.

In particular, the lake I'd been reading about directly caused the death of a lot of people. The people didn't drown, and their deaths weren't due to something obvious (like a dam bursting) but rather were from something intrinsic to the lake itself.

This is the kind of article that makes me start doing research. It's a fascinating topic, but the article itself was heavy on fluff and light on details. It had lots of exclamation points and adjectives like *amazing, stunning, deadly,* and *awesome*. There's nothing wrong with these in general; they just don't help me to understand what's really going on here. I wanted to know if this was a real thing, and if so, how could a lake actually kill so many people?

As I read, it became clear that this strange lake is nothing like what you think about when you think of a lake. When you think of a "lake," you probably have something like the following in mind (figure 3.1).

But there *are* deadly lakes. It's not a poisonous lake (that is, you can go swimming in it without any harm coming to you), and it's not a boiling lake, like this one in Yellowstone National Park.

So I had to think carefully about what such a "deadly lake" would be. Here's the research question I wrote down to get started.

Figure 3.1
A generic lake, quiet, serene, and probably not dangerous.
Credit: Daniel M. Russell

Figure 3.2
Falling into this lake at Yellowstone National Park is a bad idea.
Credit: Daniel M. Russell

Research Question 1: *Is there a lake somewhere in the world that could have somehow caused the death of nearly two thousand people?*

Here's what I did to find the answer.

To start this research, I did the simplest, fastest Google query I could think of:

[killer lake]

I know—this query seems way too obvious to work. Much to my surprise, this simple query leads you to quickly discover that **Lake Nyos**, in Cameroon, Africa, suddenly released a huge quantity of carbon dioxide (CO_2) on August 21, 1986, killing nearly two thousand people who lived nearby. CO_2 is heavier than air, so unless it's stirred up by the wind, it pools close to the ground, leading to death by asphyxiation when the bubble of CO_2 flowed silently over the nearby villages. Tragically, this happened at night, when many people were sleeping close to the ground, killing everyone and everything within a roughly 15-mile (or 24-kilometer) radius.[1]

Figure 3.3
Dead cattle near Nyos, August 1986.
© US Geological Survey

Lake Nyos is a roughly circular lake known in the geology trade as a *maar*.[2] A maar is a crater formed when a lava flow interacts violently with groundwater, blowing a giant hole in the ground. This particular maar lake is thought to have formed about four hundred years ago and is pretty deep (682 feet, or 208 meters), sitting on a porous layer of rubble, full of leftovers from earlier volcanic eruptions. CO_2 seeps upward into the water as well as into the deep lake-bottom water. There, under pressure from the water above, the CO_2 accumulates, with water pressure preventing the gas from bubbling up. It instead just dissolves into the water. In essence, this unusual configuration is creating highly pressurized soda water.

Research Question 2: *How could such a thing happen? What would make it deadly?*

If the lake were in a different place, seasonal temperature swings would mix the waters, preventing CO_2 buildup. Cold weather causes surface waters to become dense and sink, displacing lower layers upward; in spring, the process reverses. But for equatorial lakes such as Nyos and nearby Lake Monoun, these deep layers seldom mix with the top layers. These waters might be unmixed for hundreds of years.

On that night in 1986, though, something happened to suddenly cause the waters to mix, bringing the carbonated water from the bottom to the top. One theory is that boulders crashing into the lake set it off. (There were signs of a recent rockslide nearby.) Or it could have been a wind-generated wave that hit the surface just wrong, causing a mixing of the layers. As the bottom layers of water saturated with CO_2 started coming to the top, dissolved CO_2 bubbled out of the solution, and the bubbles drew larger and larger until the lake exploded like a huge shaken seltzer bottle. This explosion also brought up iron-rich water, which oxidized at the surface and turned the lake red.

In the process of reading about this, I realized that the US Geological Survey (USGS) probably had a report or two on this that would give me the level of detail I wanted, so I did a search for:

[site:USGS.gov Lake Nyos]

It showed a number of great results. (Adding the **site:** at the beginning of the query means that the results will be limited to only web pages from the USGS.gov website.)

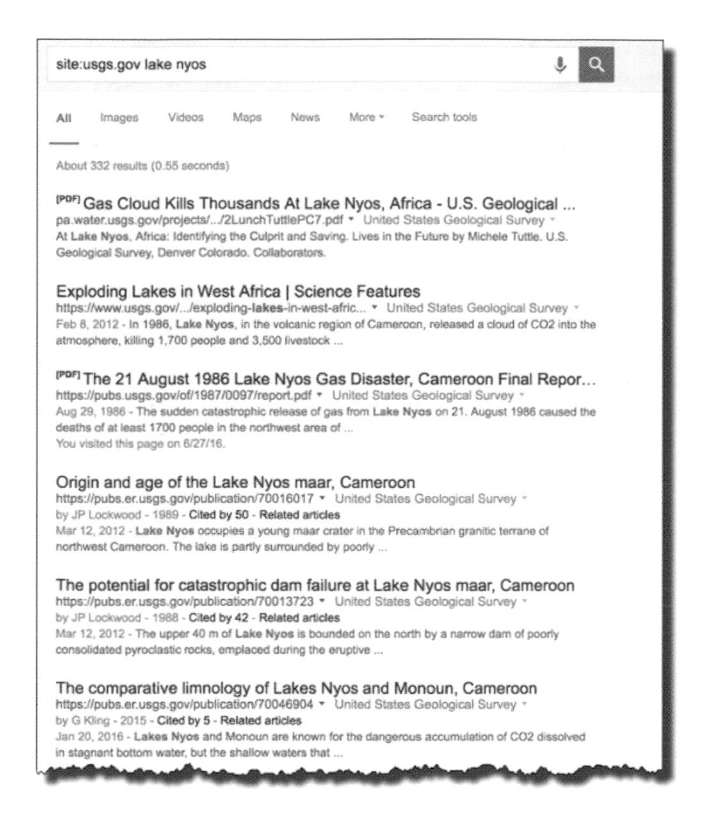

Figure 3.4
Search results for Lake Nyos that come only from USGS.gov; here I used the site: operator to limit the results to those from the USGS website.
© Google

This leads to several technical reports on the lake's CO_2 eruption. There's a USGS final report with as much detail about the eruption as a geologist might want to read. (If you're interested, I highly recommend this.)

After reading about this event for a bit, I learned that the technical term for this kind of sudden bubbling up of CO_2 from a lake is a "limnic eruption." I had no idea what a limnic eruption really is or how it works, so a quick search for this specialized term teaches me a good deal:

[limnic eruption]

Google gives me a bit more detail: a limnic eruption, "*also referred to as a lake overturn, is a rare type of natural disaster in which dissolved carbon dioxide*

Figure 3.5
Lake Nyos shoreline that's been washed away by a large wave(s).
© US Geological Survey

(CO_2) suddenly erupts from deep lake waters, forming a gas cloud that can suffocate wildlife, livestock, and humans." This is starting to make sense; I'm finding multiple sources of information that are telling me the same things.

Now what about that wave?

This search takes us to several more research reports, including ones that tell us about the wave. The *Washington Post* reported it as being 80 feet high (bit.LY/TJOS-3-3), while a report from Duke University claims that the resulting wave was a "water surge 6 m high [that] flowed over the spillway at the northern end of the lake, and a fountain of water or froth had splashed over an 80-m-high rock promontory on the southwestern shore."[3] You can see the washed-away vegetation in this picture from the USGS.

If that barren promontory is the same one that was denuded by the post–limnic eruption wave, then it likely was an 80-meter wave, and not 80 feet. (It's difficult to actually know without a better image or a topographic map to tell us the exact elevation, but if those are trees in the foreground, it well could be 80 meters.) And of course, this leads to another question.

Research Question 3: *And how high did the eruption blow water into the sky?*

To answer this question, I turned to **Google Scholar**[A] for a few reports written on the Lake Nyos eruption using the query (yes, the same query as before, but this time using just Google Scholar rather than the more general Google Web search):

[limnic eruption]

I quickly found a paper with the title "Dynamics of CO_2-Driven Lake Eruption."[4]

When you look up the author, Youxue Zhang, you'll find that he's in the University of Michigan geology department, and this paper was published in the science journal *Nature*. (Just so you'll know, it's quite hard to get things published there. So this is probably correct, at least to the best of our current knowledge.)

In this paper, Zhang creates a mathematical model of what happened in August 1986. And although he doesn't have a picture of the event on that fatal day, his reasoning is pretty compelling, and he made the following diagram to show what probably happened (figure 3.6).

This shows that the limnic eruption caused a shower of lake water roughly 300 meters into the air. (Note also that Zhang has indicated a promontory height of 80 meters.) The water would have shot up from the lake surface at a top speed of around 199 miles per hour (or 320 kilometers per hour), which is incredibly violent. You would not want to be nearby when this happens, not just because of the high velocity water jet and rain

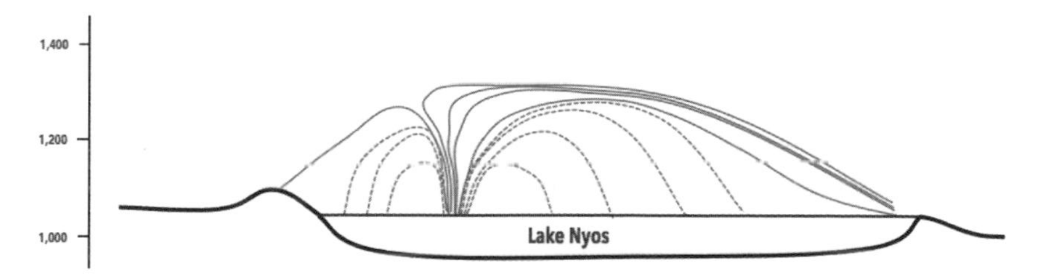

Figure 3.6
The pattern of the explosion on Lake Nyos. (Redrawn after Zhang's illustration in "Dynamics of CO_2-Driven Lake Eruptions.")
Credit: Daniel M. Russell

Figure 3.7
The degassing fountain that runs continuously, powered only by the pressure of expanding gases as the water flows from two hundred meters below to the surface of Lake Nyos.
© US Geological Survey

of carbonated water, but because that's when all the CO_2 is released into the air. It's probably a spectacular sight, but one that you'd want to watch from a safe distance.

So what's going on with Lake Nyos today? It's still dangerously full of high-pressure CO_2, but there has been an effort to depressurize the lower levels of the lake by dropping a 200-meter tube from the surface to the lower part of the lake. As you might expect, when you make a connection between the surface and the pressurized bottom of the lake, you get a fountain, which they hope is removing the pressurized CO_2 fast enough to prevent a similar disaster from happening in the future (figure 3.7).

But of course, where there's one maar waiting to turn over, there might be others. Nearby, Lake Kivu could be another such lake with a huge CO_2 reservoir just waiting for the right trigger to overturn and release its CO_2 contents into the atmosphere (based on the USGS report on this hazard). If this happens, it could easily kill many more people, as there are around two million people living nearby.

Research Lessons

There are a few lessons here.

1. *Short and obvious searches often work well.* Much to my surprise, a query like [killer lake] actually led instantly to Lake Nyos. This startled me; there really haven't been other killer lakes? Apparently not. The biggest lesson I pick up from this is that obvious queries work—but only if you have some kind of target in mind. (That is, [killer lake] isn't the kind of query any normal person would do unless they were looking for something like the story behind Lake Nyos.)

2. *Checking a known expert site (e.g., USGS.gov) is a great strategy for getting in-depth articles.* While it's easy to find lots of articles on Lake Nyos, scholarly ones are best when they come from a known source that does a lot of work in this area. In this case, I happened to know about the USGS in advance, so I just restricted my search to the articles and papers inside USGS.gov by using the **site:** operator. (This is probably the single operator that I use more than any other to limit search results to just those from one site.)

3. *Specialized terms, when you learn them, are great!* This is a lesson we've seen a bunch; when you see these special terms (or phrases, such as limnic eruption), use them for your search terms. (But be sure you know exactly what they mean or you might end up doing some time-wasting searches. And be sure to get your spelling right: a search for a **limbus eruption** will tell you a lot about diseases of the eye, and nothing about savage geology; that would be a **limnic eruption**.)

4. *Triangulating multiple sources is a requirement when you're doing research.* As you see, there are reports of the tsunami being eighty feet … or eighty meters. That's a huge difference. When you're doing research like this, keep looking around; you might find different data (or points of view) around a topic—even things that should be easy to verify.

5. *Remember that Google Scholar is a great collection of high-quality articles and papers that anyone can access for free.* While it's a collection of academic and scientific papers, it's NOT just for academics but rather for everyone to use. Google Scholar gives everyone equal access to a rich source of knowledge that's otherwise difficult to find. The kind of knowledge that you can discover in Google Scholar is comprehensive and authoritative, which makes it perfect for you to use when studying lakes that explode due to subterranean

CO_2 sources. If you're researching something that's a bit technical, do a search for your topic in Google Scholar as a way to get access to some of the best research and information available.

How to Do It

A. Google Scholar. Google offers Scholar.Google.com as a collection of peer-reviewed scientific papers from journals, magazines, and technical conferences. It's simple to use; just visit the site and do a search for the topic of your choice. For instance, the search I did above, [limnic eruption], gives different results on "regular" Google versus Google Scholar. The results from Google Scholar are going to be much more detailed and technical than those from Google.com; you won't see any videos in Google Scholar results, but you will find authoritative articles written by geologists and research scientists from around the world.

Try This Yourself

While the USGS is the keeper of much of geologic data about the United States (and other places in the world as well), the British Geological Survey is the equivalent governmental organization in Britain. Its website is www.bgs.ac.uk. Now that you know its web address, you can use some of the key lessons of this chapter to answer questions that the British Geological Survey would have deep knowledge about. For instance, Stonehenge, the Paleolithic monument in Wiltshire, England, has many mysteries. (See figure 3.8.) One of these is how well known it was throughout the prehistoric world. Did people in other parts of Europe visit the site? For example, did people from the Mediterranean come to visit and maybe even stay at Stonehenge?

How would you make a query to find out?

Figure 3.8
Stonehenge has been around for more than four thousand years, yet it still has many unanswered questions. Can you find out if the British Geological Survey has reason to believe that people from the Mediterranean visited the site?
© Pixabay; rights granted for commercial use

4 Things You Notice While Traveling: How and When to Switch Search Modes to Find Information

Satisfying your curiosity while on the road is usually a quick search on your mobile device. Here are some fast ways to get to the answers that you seek.

I don't know about you, but when I'm traveling I often end up with small questions that arise during the day from reading the local news, listening to the radio on those long drives, or watching slightly sketchy TV in a random hotel room.

Mostly, these questions that pop up are fairly easy to answer, but they're like mosquito bites—you're not really satisfied until you know the answers. Here are a few questions that came up for me in my travels over the past couple of months. As they happen, I just jot them down and look these things up when I get home. How many of these have you also had?

Research Question 1: *You're driving a rental car into the gas station. Is there some way to tell (without getting out of the car) which side the gas cap is on? Is it on the left or right?*

It's really annoying to have to get out of the car, walk around, and discover it.

The answer is obvious, once you know. But I had to look it up; more important, I had to realize that this was a problem, a minor one to be sure, yet it's a problem for which I could probably search out the answer. Here was my query: [car dashboard fuel side] (see figure 4.1).

Interestingly, this also led me to a Snopes.com article asking if the gas cap side was indicated by which side the pump handle on the ICON was

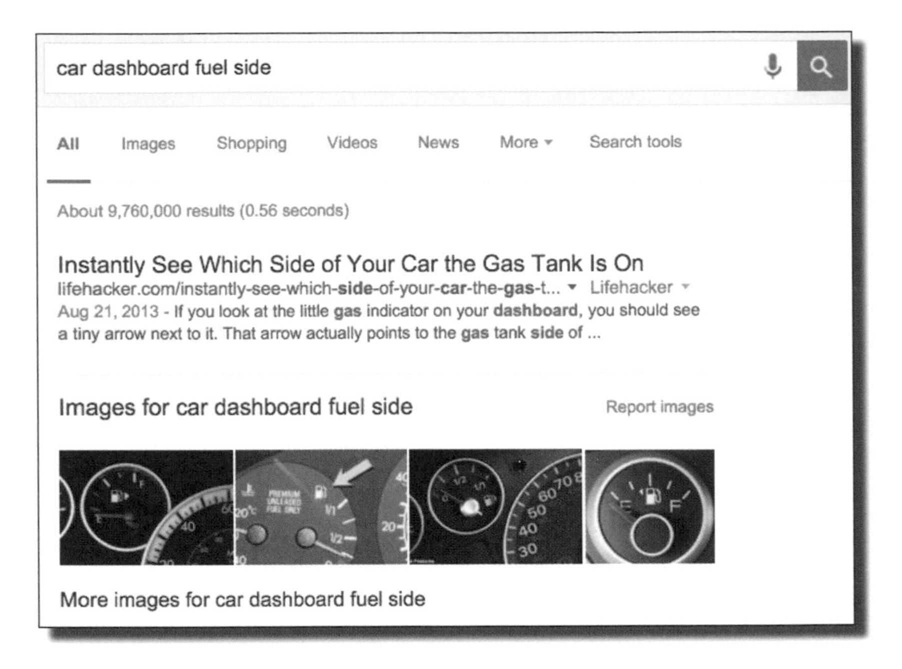

Figure 4.1
A quick query about fuel indicators on the dashboard uses just the essential query terms.
Credit: Google and the Google logo are registered trademarks of Google Inc., used with permission

on.[1] I hadn't thought of that interpretation. But no, it's the triangle that points to the correct side of the car; the pump handle doesn't actually indicate anything.

By the way, what's the correct term for that part of the car where the gas cap is located? What's it called? I did the following query:

[gas cap diagram]

I ended up finding out that the gas cap is the cover (the cap) on the "fuel filler hose" that leads to the fuel tank. Another term I didn't know! In the process, I learned that this is the term you should use when looking up information about the "gas cap" in the booklet that came along with the car. The language that you and I might use to talk about gas caps isn't necessarily the same as the language that people who *write* the manuals use.

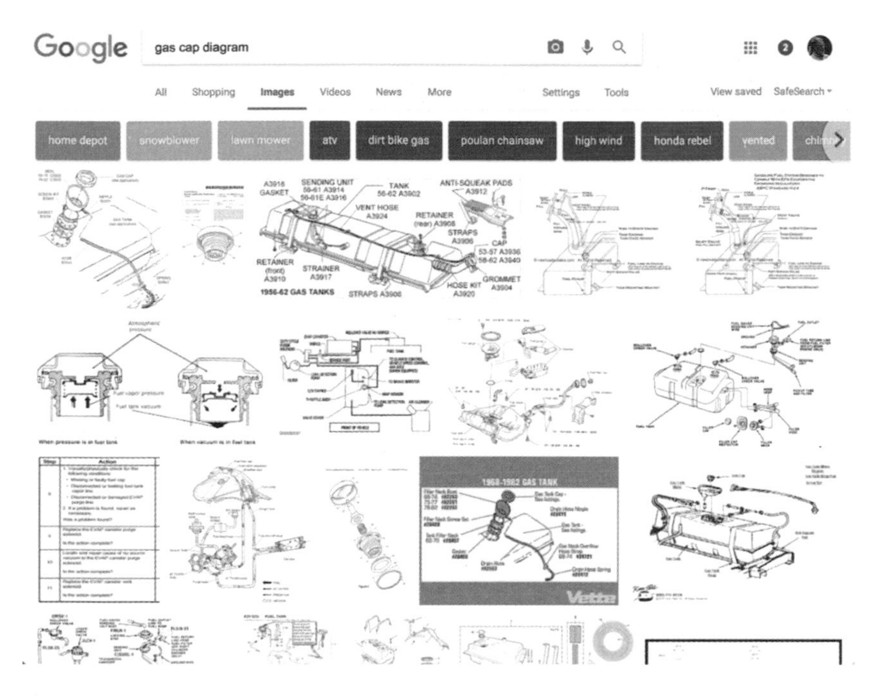

Figure 4.2
The query [gas cap diagram] gives you a set of results that are all diagrams that have a gas cap in them somewhere. Adding the extra context term "diagram" is a handy search tip to know when you want to learn how things are built, organized, or laid out.
Credit: Google and the Google logo are registered trademarks of Google Inc., used with permission

It's also worth remembering this search trick of adding in a term that describes what you're looking for; I call this a *context term* because it provides a bit more context for the kind of query results you want.

With this query, if you look in Google Images (which is probably where you'd expect to find the diagrams), you'll find lots of useful diagrams showing you how the fuel filler hose is located, and where the cover is located.

Research Question 2: *I saw Mount Rushmore, and while I was being amazed, I also wondered about how it was designed by the sculptor. What was the original design for Mount Rushmore? Was what we see today what he originally thought it would be? How different was the original design?*

Figure 4.3
An early design had presidents shown to their waist rather than just their heads. But you stop sculpting when you run out of funding, as the project did in 1939.
Credit: Library of Congress; © Rise Studio

My first query was intended to find the original design for the monument:

[Mount Rushmore original design]

It led to a number of sites, all of which pointed out that the original design was as shown in the photograph in figure 4.3. Of course, it was also going to be at a different location (at the rocky outcrop called Needles, not far away from where Mount Rushmore is), and Thomas Jefferson was originally supposed to be at George Washington's left, until they found that the rock there wasn't suited for sculpting.

With any project of this scale, there are many designs in the ramp up to actually getting it running. Another approach to answering this question would be to expect that there would be multiple models that were developed along the way and then do a query like this:

[Mount Rushmore models]

This gives a set of results showing that Gutzon Borglum (the sculptor) had many variations on the design of Mount Rushmore—so many that it's

Figure 4.4
The way Mount Rushmore looks today is quite different than the early designs.
Credit: Winkelvi, Wikipedia, shared under a CC BY 4.0 License

difficult to identify just one as the "original." The differences between the variations are fairly substantial, although the project was always going to be presidential.

Research Question 3: *How do you know which is the "right" side of a towel? (Yes, there's a right and wrong side. Have you been doing it wrong all these years?)*

Yeah, who knew? I saw this while traveling in a seaside community. There was an ad for a special kind of beach towel that touted its especially wonderful "right side." What? How could a towel have a wrong side? That made me wonder, *IS there a right and wrong side to a towel?*

This was a quick and simple query; [towel right wrong side] found many articles telling me that the side with the long loops (and therefore more surface area) is the side meant for drying. The side with the short loops is meant for everything else (maybe sitting on the beach?). As with many things like this, use the side you like.

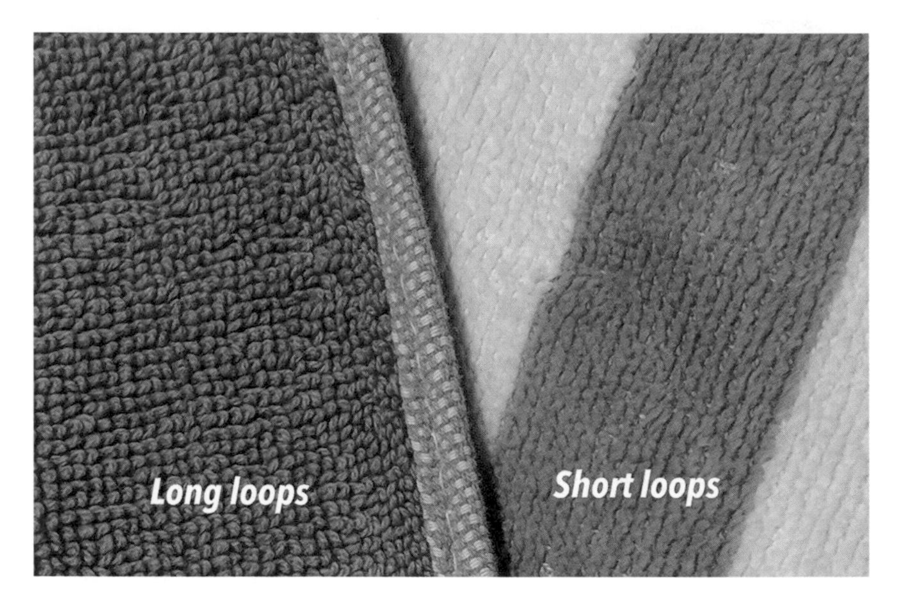

Figure 4.5
Most towels have long loops on one side, and short loops on the other. The side with long loops is somewhat more absorbent, making it the *right side* for drying off.
Credit: Daniel M. Russell

Research Question 4: *People keep saying things, but I don't know what they mean! For instance, just today, someone said, "It was taken care of in one fell swoop." What does "one fell swoop" mean?*

I'd start this piece of research with the straightforward:

[one fell swoop]

Again, the search is easy, but finding the answer at PhraseFinder tells me that it means "a sudden action" and comes from William Shakespeare's play *Macbeth.*[2] The original use was in this bit of dialogue:

MACDUFF: (on hearing that his family and servants have all been killed)

All my pretty ones?

Did you say all? O hell-kite! All?

What, all my pretty chickens and their dam

At one fell swoop?

Here, Shakespeare uses the imagery of a hunting bird (in this case, a "hell-kite"; you could go look it up!) that did a "fell swoop" to suggest the animal and savage nature of the deadly attack on Macduff's family by Macbeth's assassins.

The "swoop" is when a raptor (such as the "kite" mentioned in line two) makes a sudden, rapid descent from on high, usually called a "stoop" today. The stoop is the way that raptors stun and kill their prey. They literally "swoop" down on them. "Fell" as an adjective here just means "evil" or "fierce." Today this phrase has come to mean "a sudden stroke, action, or attack."

Research Question 5: *At a small country store, I heard someone say, "Oh, heavens to murgatroyd!" Really? What does that even mean? And who said this originally anyway?*

You have to be a bit cautious when searching for etymologies (the origin of words and phrases); in general, they're complicated. But as a first guess, my search was:

[heavens to murgatroid]

Notice that I left the "oh" out of the phrase. When you're searching for quotes or phrases, usually the best way is to search for just the core of the phrase—that is, the part that's probably going to be preserved over multiple tellings and repetitions. I also did NOT put it in quotes because I figured that this sequence of words would be idiosyncratic enough to work.

I got the following search result (figure 4.6).

The search is quietly telling me that my search is actually misspelled. Who knew? The *correct* spelling is listed in the helpful phrase, "Showing results for murgatroyd." That wasn't what I originally put in my query, which was "murgatroid." (If I was convinced that my spelling was correct, I could put that **single word in double quotes**[A], which would then turn off spell correcting.

Looking at these results, we can see that at very least, it was said by the cartoon character Snagglepuss.

Out of sheer curiosity, I clicked on several of the top ten results, including the Wikipedia entry that directed me back to PhraseFinder. It turns out that our friends at PhraseFinder have a story for this too. They claim that "heavens to murgatroyd" was first uttered by Bert Lahr (an actor best

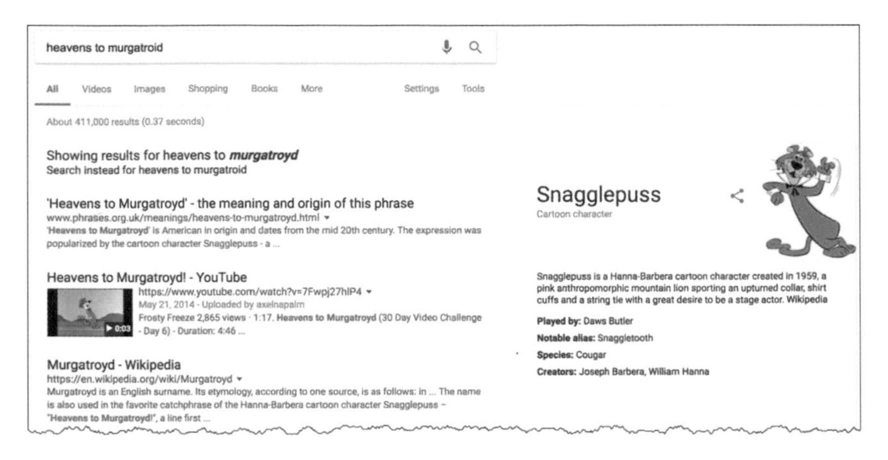

Figure 4.6
Searching for a common phrase works well, even if you misspell it.
Credit: Google and the Google logo are registered trademarks of Google Inc., used with permission

known as the Cowardly Lion in the *Wizard of Oz* movie) in the 1944 film *Meet the People*. Dick Powell and Lucille Ball star in this World War II era musical revue. Ball plays a popular but stuck-up Broadway star who leaves the bright lights to become a welder in a shipyard as a way to contribute to the war effort. Here she meets and falls in love with coworker Powell. Since this is a wartime musical, the plotline is periodically abandoned for guest star cameos of famous actors (of the time) Virginia O'Brien and Lahr, and musical groups like the Vaughn Monroe Orchestra or Spike Jones and His City Slickers.

While that was the phrase's first appearance, it was popularized by the 1959 Hanna-Barbera cartoon character Snagglepuss, a pink mountain lion with grand aspirations to be an actor. Interestingly, the voice of Snagglepuss was originally by actor Dawes Butler, who obviously imitated the voice style of Lahr playing the Cowardly Lion. That voice style was so similar that Lahr sued Kellogg's when it used Snagglepuss in some of its TV commercials. Lahr was concerned that viewers might think that he was endorsing the cereal. As part of the settlement, subsequent commercials with Snagglepuss had to have a credit line saying, "Snagglepuss voice by Dawes Butler," making Butler quite possibly the only voice actor to get a credit line in a commercial!

Now that we've found that this is a common exclamation by the character of Snagglepuss, watching one or two videos quickly tells us that it's a general phrase of exasperation, a bit like "oh heavens to Betsy." Of course, I now wonder who Betsy was—but I'll leave that for you to find out on your own.

Research Question 6: *In the story of David and Goliath, David kills the giant with a single stone from his sling. Is this really possible? What kind of a slingshot could do that?*

A simple query, [David sling stone] or [David slingshot], leads to a wealth of information about the "traditional sling" (as opposed to the surgical tube or heavy rubber band version).

A traditional sling is just a length of twine (or leather) with a small pocket into which you can place a stone or small projectile. David's traditional sling would have looked like the following (no rubber bands needed).

How powerful can such a sling be? Isn't this just a kid's plaything?

Figure 4.7

The classical sling from biblical times is made of string and a woven pocket for the stone—simple, effective, and deadly.

© Yair Aronshtam

The sling is simple to make and yet much more powerful than you'd think. When I was in high school, I read an interesting article in *Scientific American* on "The Sling as a Weapon" (1973).[3] Being a teen, I made one and proceeded to test it out. I chose a smooth stone about one inch (two centimeters) in diameter, whirled it around my head a couple of times, and let it fly. The stone hit the side of our wood-shingle-sided garage and went THROUGH the wall!

Fascinated, I quickly found that I could easily put such a stone through a quarter-inch plywood at twenty paces. (I quickly stopped testing in my backyard because I realized a missed shot could easily break windows and potentially really hurt someone. This simple sling has enormous power. I soon came to believe that it could kill someone, including Goliath-sized giants.)

On a trip to the seashore, I ended up slinging stones out into the ocean, and got them going fast and far enough that I'd often lose sight of them before they hit the water. From my perspective, they'd just disappear into the distance.

One of the best sources I found is the website Slinging.org (its tagline is "Stone Age ballistics"), which has an extensive list of publications about slings and their use as well as a lot of practical advice.

While there is a great debate about whether or not the stone actually killed Goliath by striking him in the forehead (wasn't he wearing a helmet?), there's no doubt that a sling could seriously injure and quite possibly kill someone.

That first query led me to a lot of results, so I found myself scanning through them, trying to pick out the articles that would tell me more about the uses of slings and stones as weapons. One of those articles was a rather-promising one by Eric Skov that appeared in a 2011 issue of *Nebraska Anthropologist* titled "Sling Technology: Towards an Understanding of Capabilities."[4]

I read fairly widely, but even I'm not in the habit of reading back issues of the *Nebraska Anthropologist*, yet after I read this article, I might have to start.

In this article on the devastating power of the sling, Skov shows pretty conclusively that a relatively small lead bullet flung by a sling could easily penetrate the skull, allowing even a relatively small shepherd boy to take down a giant. David and Goliath indeed. (Of course, if you're interested in what seems like an uneven fight, I have to recommend Malcolm Gladwell's book *David and Goliath: Underdogs, Misfits, and the Art of Battling Giants*, if only for his telling of the David and Goliath story, emphasizing that David was a smart kid who knew how powerful a sling in a shepherd's skilled hand could be.)[5]

To this day, slings are still being used in the Middle East, where they're powerful weapons in the hands of people who have used them for thousands of years.

Research Lessons

The real lesson here is obvious.

1. *Look up stuff that's interesting, unknown, or unclear*. Learn to recognize the little annoyances that you can reframe as a small research question. In my case, the hassle of getting out and looking for the darn gas cap was annoying enough to make me wonder, "Has anyone else had this problem?" Hence my query above. And hence all these queries that came from this. None of them are difficult, so it really shows how much great content is available, if only you go and look for it!

2. *Adding a context term (like* ***diagram****) to your search can tailor the results toward useful information sources*. Adding in a *context term* is something that expert searchers do when they want to limit the search results to a particular kind of result. Other useful context terms also describe the kinds of results that you might want to see. For instance, adding the term *album* (when looking for a collection of images on a topic), *glossary* (to find collections of words on a particular topic), or *overview* (when you're looking for a short summary on a given subject).

3. *Putting a single word in double quotes makes Google search for that word without any spell corrections or substitutions*. This is useful when you're looking for an intentionally misspelled word (for instance, when you're searching for a word that's spelled as part of a dialect, such as "dat" for "that"), or a place-name with a spelling that's close to a regular word (such as "Noble," which is close to "Nobel" yet different at the same time).

How to Do It

A. Single word in double quotes. It might seem odd to do a search with a single word in double quotes (like the following).

There ARE times when you want to override the Google spelling corrector and use the search term the way you specified it. In this case, Darwwen is a friend with an unusual spelling of his name (figure 4.8). In cases like that, you can put double quotes around the search term to mean *search for exactly this term*. (You are implicitly telling Google, "Please don't help out by fixing the spelling for me; this is really what I mean.")

Of course, most of the time the spelling correction is what you want. But this gives you a way to get around that system if you need to do so. (As in this example, it's most often for unusual spellings of names.)

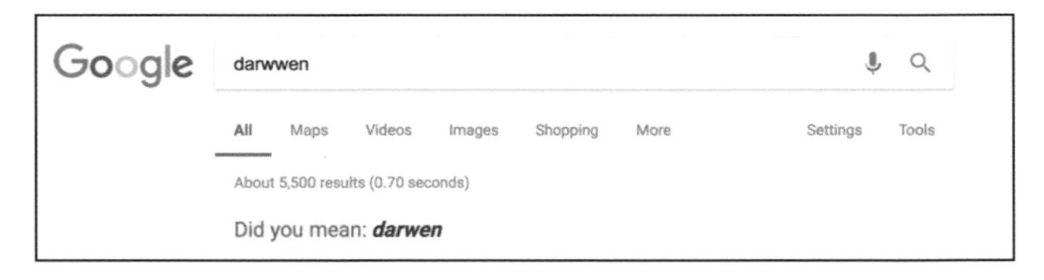

Figure 4.8
Double quoting a single word will return results with *exactly* that word. There's a suggestion for a more standard spelling, but you can ignore that and see only the results with this (slightly unusual) name.
Credit: Google and the Google logo are registered trademarks of Google Inc., used with permission

Try This Yourself

This chapter is really about how to quickly satisfy your curiosity when you see something that piques your interest. When this happens to you, you needn't just wonder about it; now you can go do a bit of research to find out the backstory.

For example, the other day while driving down a country road in Arizona, I noticed a series of small signs that rhymed when read aloud. This series of signs read:

If daisies are
your favorite flower
keep pushing up
those miles per hour!

People have told me that this was once a common sight on roads in the United States, but no longer.

Who put up these signs, and why? What was the point? (Think: How would you search for something like this playful poetry?)

5 Is That Plant Poisonous or Not? How to Find Highly Localized and Domain-Specific Information

Is that plant safe to eat? How would you find out in a hurry?

Many parents have had this awful moment: your kid comes in from playing outside, and their mouth has little green leaves, petals, or white sap from something she's been eating. The thing you need to figure out—*quickly!*—is whether it's poisonous or not.

Here's my story about trying to quickly identify a local flower.

This time of year, a particular kind of small, white flower blooms all over the valley where I live in the San Francisco Bay Area of California. You see it everywhere—in open fields, along roadsides, anyplace where there's not a lot of water yet a lot of relatively untended land.

Imagine yourself the parent of a child who comes in carrying this flower, with one-half eaten hanging from her mouth. What's worse, it smells terrible.

What should you do?

The research question here is fairly simple and obvious.

Research Question 1: *What is this plant?*

Research Question 2: *Is it poisonous or not?*

As you can probably tell, I enjoy flowers—especially the local wildflowers that appear everywhere in the valley and woodlands near my home.

So when I saw this particular flower in my daughter's hand, I recognized it instantly as a member of the *Compositae* family. That's the family name

Figure 5.1
A mysterious flower appears in your daughter's mouth. Should you be worried?
Credit: Daniel M. Russell

for flowers that look like this: they all have multiple flowers in a single disk (botanically "composite"), usually with petals pointing straight out from the center. This family includes sunflowers, dandelions, asters, and similar flowers.

You might not recognize this as a member of the *Compositae* family, so maybe you'd like to call Poison Control first. The quick way to find out what number to call is with the query:

[poison control]

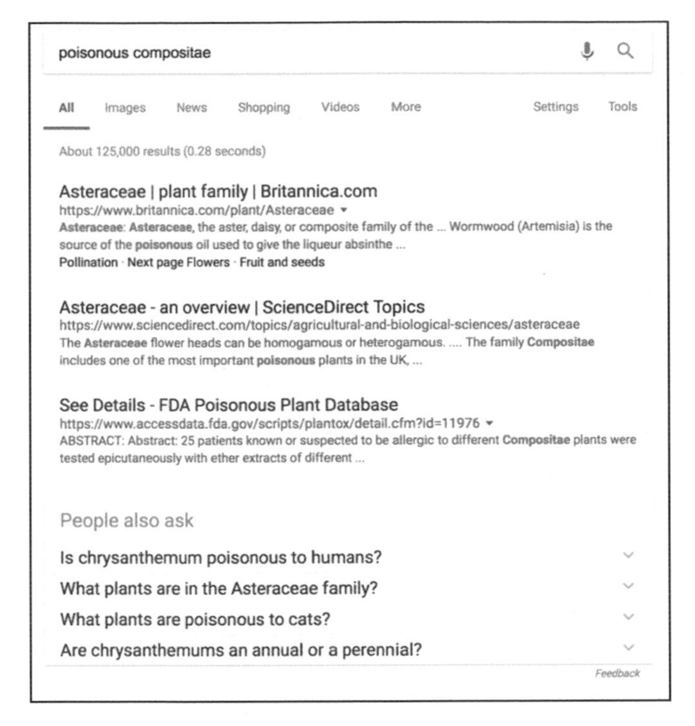

Figure 5.2
A quick check to see if this kind of *Compositae* plant is generally poisonous.
Credit: Google and the Google logo are registered trademarks of Google Inc., used with permission

That gives you the phone number of the emergency service that can help you out. (This is a handy trick to know. It works no matter where you are.)

Poison Control, however, might not know what your local plants are, so while you're on hold, you could look up this information.

But (this is a key point) many *Compositae* flowers and plants are edible—both the leaves, stems, and flowers. Some common edible composite flowering plants are *Lactuca sativa* (lettuce), *Cichorium* (chicory), *Cynara scolymus* (globe artichoke), *Helianthus annuus* (sunflower), *Carthamus tinctorius* (safflower), and *Helianthus tuberosus* (Jerusalem artichoke).

So knowing that it's a *Compositae,* I did a quick search to see if these flowers are, in general, bad for your health with the query:

[poisonous compositae]

That gives me the following search results page (the SERP in figure 5.2).

Figure 5.3
The mystery flower growing by the side of the road, looking a bit weedy.
Credit: Daniel M. Russell

Reading this SERP tells me a couple of things right off the bat. First, looking at the bolded term *Asteraceae* in the first result tells me that my word *Compositae* is probably an outmoded or, at the very least, slightly alternative name. (What can I say? I learned my botany in a different era.)

The link to the Food and Drug Administration's Poisonous Plant Database is worrisome, but clicking on the link takes me to a slightly scary list of noxious plants.[1] Fortunately, scanning that page lets me know that its list of *Asteraceae/Compositae* plants is all about *contact allergies*. That is, the plants are not poisonous but are instead just irritating.

Yet as the third question in the "People Also Ask" suggests, chrysanthemums ARE toxic to cats (and dogs)

So there are a few *Compositae* that are somewhat poisonous in large amounts. The most common is probably burdock, *Arctium* genus, leaves, which can cause a rash. But even then, burdock roots are good to eat, so it's not really toxic per se—just really irritating. In the United Kingdom, ragwort, *Senecio jacobaea*, is poisonous, but we're not near the United Kingdom; we're in California.

A little background: the name *Asteraceae* comes from *Aster*, the most prominent genus in the family, and derives from the Greek word meaning star. The flower has a clear star shape. *Compositae* refers to the fact that the family is one of the few flowering plants that have a bunch of tiny flowers growing tightly side by side in a center.

So at this point, I'm not especially worried. Our challenge isn't to feel good about our background knowledge, though, but rather to identify what the kid is holding in her hand. We want to figure out exactly what it is. If your kid got into a patch of flowers like this one and started munching, should you be worried?

In my case, I'd just turn to one of my *plant identification guides* such as *A Field Guide to Pacific States Wildflowers*.[2] These books all have a "key"—that is, a linked set of questions about the flower that lead to an identification.

The questions in the key are like this. They're short, use a fair bit of technical language, and lead you to a plant family or species name. Here's the last question you have to answer before making a positive identification of a flower as a *Compositae*:

> 87a. Flowers in racemes, spikes, or solitary. …
> 87b. Numerous flowers in heads, sunflower-like. Each flower tubular (some with outer ray flowers). **Compositae.**

If the flower isn't in *racemes, spikes, or solitary*, it's either a *Caprifoliaceae* (and pink) or *Campanulaceae* (and blue). In essence, if you answer each question correctly and then follow the numbers, you'll end up at the right flower identification. It's basically a flowchart for identifying flowers.

But it's a little tricky. In practice, I end up looking up every other word (what's a "raceme"? What's an "outer ray flower"?), and going up and down the tree of questions until I finally get to one that matches all the test questions.

We don't have time for that here. So what can we online searchers do to help?

There are a couple of ways to do this.

A. Find an online wildflower identification book. Typically, these cost some money. As you can imagine, there aren't a lot of free ones available online. There are online versions of various field guides that you can buy (say, on iBooks, Amazon, or Google Play Books). You could then just use them to ID the flower in the normal way by working your way up and down the key questions. This is a great use case for tablets or large phones with a reasonably sized display. If it's an e-book, you can carry lots of field guides with you; this is especially handy as they tend to be big and bulky. The e-versions have links and bookmarks to help you flip back and forth.

B. Search for online wildflower guides specific to the area. This is the approach I took. Knowing that there are a LOT of people (besides me) who want to identify flowers, I did a simple search:

[wildflowers San Francisco Bay Area]

This query brings up a number of resources that seem like they might work. Notice that I added "San Francisco Bay Area" in the query. Even though Google would probably assume I meant *local* flowers, I wanted to be sure that I wasn't also looking at flowers from the far north or far south of California.

After looking through a number of wildflower collections, I finally find two photographers' collection of white flowers, with a flower that looks similar to the one in the photo above. In their website collection, John Raithel and Linda Herbert label it as **Hayfield tarweed** (*Hemizonia congesta*), and in their photo, it looks a LOT like the flower above (figure 5.4).

Getting to this point took a little time, as I had to look through a few collections of white wildflowers before I found this image. As nice as it is, this little picture isn't quite large enough for me to be sure that the identification of my daughter's flower is quite right. So I did another search for the Latin name given on the web page, *Hemizonia congesta*. The obvious search query at this point is to find out more about the biological name.

Figure 5.4
After scanning several online collections of San Francisco wildflowers, I finally found this one—Hayfield tarweed (*Hemizonia congesta*).
Credit: www.rahul.net/raithel/nature/

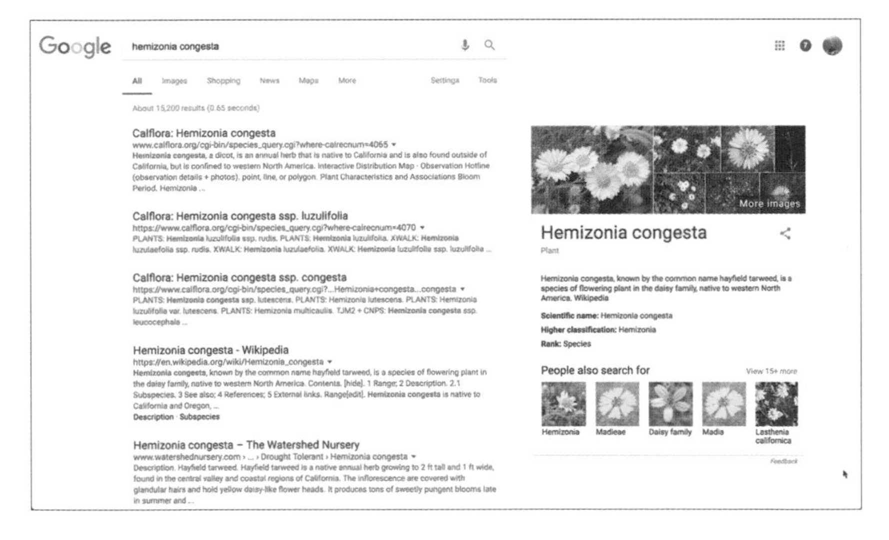

Figure 5.5
A simple search for the Latin name of the flower often gives great results.
Credit: Google and the Google logo are registered trademarks of Google Inc., used with permission

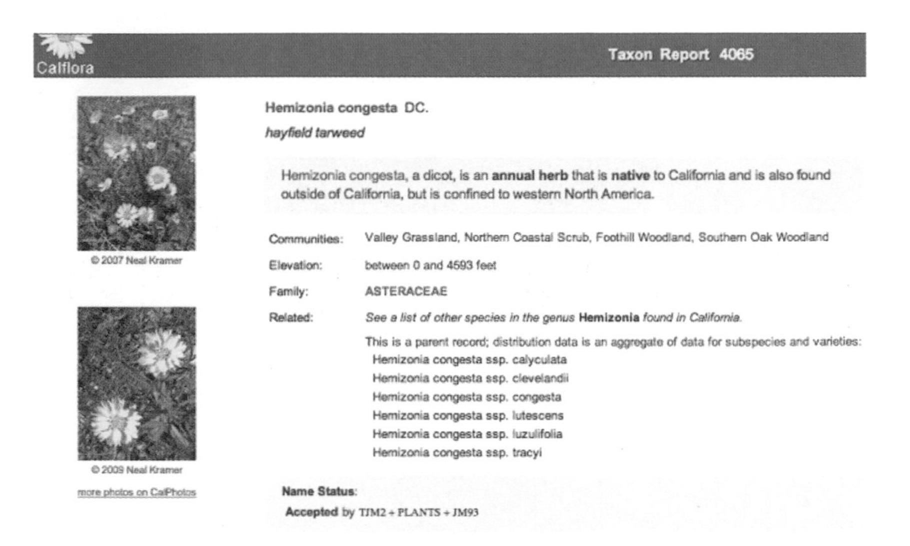

Figure 5.6
The Calflora site is a large digital library site of information about many of the flowers in California. In this page about *Hemizonia congesta*, the Hayfield tarweed, you can see pictures of the yellow subspecies. But what about the white version?
Credit: Neal Kramer © 2007, The Calflora Database [a nonprofit organization], http://www.calflora.org

As you can see in figure 5.5, the top three results are from the Calflora.org website. I know that Calflora.org is a massive, high-quality database of flowers and plants in California that was originally started by people in the US Forest Service, and who also work with the botanists at the University of California at Berkeley. (You can learn this by checking the "history" tab on the Calflora site, or you could search just for information about the organization by doing a search for [Calflora] and reading about its mission.) Bottom line: for matters botanical, especially in California, I tend to believe Calflora; it's maintained by botanists, and they're careful with their content.

I start to get worried, though, when I check the Calflora site. When I look at the page for *Hemizonia congesta*, nearly all the flowers are yellow. Uh oh. They *look* like the same flower, but they're not white. Is this NOT the right flower?

See those links to *Hemizonia congesta ssp*? Clicking on those shows many photos of this flower from different "ssp" (that is, "subspecies"; you can

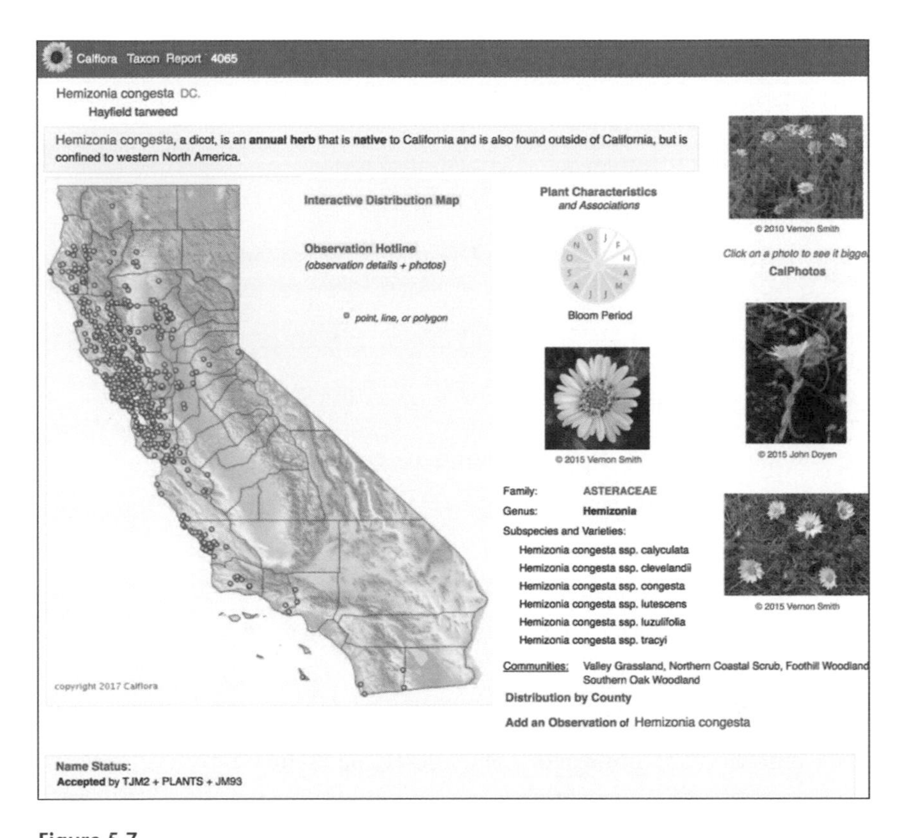

Figure 5.7

Hemizonia congesta congesta is the white subspecies of the Hayfield tarweed.
Credit: https://www.calflora.org/cgi-bin/species_query.cgi?where-calrecnum=4065; Vernon Smith, PhD, and Jack Doyen, 2015; © Calflora

find this out by searching for [define ssp botany] to learn what that specialty term means). It turns out that there are lots of subspecies, with color variation between yellow and white as part of the defining characteristic. And it turns out that *Hemizonia congesta congesta* (yes, with the word *congesta* repeated twice) is the white version of the plant.

Whew! So it IS *Hemizonia congesta*, and most likely the subspecies known as *congesta*, which is white, not *lutescens*, which is yellow.

After a little digging around in the intricacies of the naming of this little flower, you'll find a quiet controversy about whether or not the various subspecies of tarweed should have different names, what that name should be, and how people agree to give it a particular scientific name. This doesn't matter much to me now, but as we saw at the beginning with the *Compositae* versus *Asteraeae* difference, even scientists sometimes change their minds about what to call things.[3]

Now we have the answer to the first question.

1. What is this plant?

Answer: *Hemizonia congesta*, the "Hayfield tarweed" in a white subspecies variation.

Lets now return to the second research question.

2. Is it poisonous or not?

To find out, I did two quick searches for:

[tarweed edibility]

and

[tarweed poisonous]

The answers come back quickly. Not only is it NOT poisonous; the local Indians would eat the seeds, aggressively harvesting them in the fall to make pinole (a form of meal or watery gruel, made from ground-up acorns, seeds, and wild grain). Reading a bit more, I discovered that they also would burn over the fields to ensure good growing conditions for next year. You wouldn't go to all this trouble if the plant were really poisonous!

This makes me think, *How was tarweed used in early California?* As we've seen, it's clear that tarweed was sought out and harvested for food. So I did two searches to follow up on this.

[tarweed Indians]
[tarweed harvesting]

Why is that query [tarweed Indians]? (As opposed to the more politically correct terms "Native Americans" or "indigenous people"?)

Short answer: because that's where the content is. It's useful to sometimes use older terms when searching. (It's fairly easy to compare [tarweed Indians] and [tarweed Native Americans], say, and quickly discover that "Indians" is a much more common term and gives us much better results!

But in both cases, I was able to find many references to the harvesting and preparation of the seeds for food.

Figure 5.8
The *Tarweed Gatherer* by Grace Carpenter Hudson (1865–1937), used with permission of the Grace Hudson Museum and Sun House, city of Ukiah, California.
Credit: Collection of the Grace Hudson Museum and Sun House, city of Ukiah, California, www.gracehudsonmuseum.org

One of the loveliest references I could find is a link to this image of a Pomo woman kitted out to collect tarweed. (Although from the title, she's apparently ready to collect the seeds of the *Madia elegans* flowers, also called tarweed, which look, taste, and smell much like *Hemizonia*. They're all called tarweed, which is a nice example of why the Latin botanical name is much more specific and useful for searching.)

The woman pictured is well-known basket maker Joseppa Pinto Dick (born circa 1860 and died 1905), who was Yokayo Pomo, a native community southeast of Ukiah in Northern California. Here she is ready to collect the plentiful seeds of the *Madia elegans* flower, a near relative of our Hayfield tarweed, *Hemizonia congesta congesta*.

Not only is the common tarweed not poisonous; but it's edible and considered quite good by those who grew up with it. I tried a couple of seeds, and found it hard to get past their resinous flavor. They were a bit too turpentine tasting for me. But perhaps people who like retsina (the pine-flavored Greek wine) would like tarweed as well.

So if you pick one or even walk through a field of them, you'll be wearing that smell on your hands and clothes until you can scrub it clear. It's an intense smell—one that reminds you happily of late summer in the golden fields of California or a stink that you can't wait to get away from. It's the cilantro of summery scents—appealing to some, and despised by others. One writer described the tarweed aroma as "pungent, soapy, with an intriguing undertone of kerosene."[4] She was right.

Research Lessons

1. *When you have a specific name for something, use it.* In this case, we found the scientific, binomial, Latin name to uniquely identify our mystery flower.

2. *Likewise, when you've found a specialized database (such as Calflora) that's dedicated to a single topic (such as the flowers of California), use it; it probably has great high-quality data.*

3. *Names change—even within the scientific community.* From the start we learned that there are two names for this family of flowers. The name *Asteraceae* comes from the genus *Aster*, from the Greek word meaning star, referring to the starlike shape of the flowers. Meanwhile, *Compositae* is an older name that refers to the fact that flowers in this family are composites, like a daisy, dandelion, or sunflower. (Pay attention, and look at chapter 11 about changing terminology.)

4. *Sometimes an online search means manually searching through collections.* When searching for the white flower based on the picture, I really knew fairly little about the flower other than how it looked. So there just wasn't much to go on, and I had to look one picture at a time through a number of collected photo albums. Sure, it felt a little like looking for a criminal in a lineup, but it didn't take long to spot the suspect in the online photo books.

5. *A "key" is a specialized kind of flowchart for identifying flowers (or animals, fossils, kinds of pollen, or microorganisms).* When you really want to learn how to identify flowers, you will at some point have to learn how to use a flower key, which leads you through a sequence of questions about the flower in question. In the process, you'll learn a lot about botany in order to understand the questions. ("Flowers in racemes, spikes, or solitary?")

Try This Yourself

A useful tactic when searching for specific information is to use specific terms. This might seem obvious (after all, a search engine uses the information you give it), but one of the most common search mistakes that I see people making is to use fairly generic language, hoping that Google will guess what they mean. The remarkable thing is how often this actually works.

Yet if you can use a specific term for your search, it will increase your accuracy.

For example, you might be able to search for a pain in your leg by doing a query for:

[pain in leg]

But then you'd spend some time reading through all the hits trying to determine which of them describes the pain you feel in *that* part of your leg. Instead, if you can use the right word for the pain in the muscle that's just to the side of your shin bone, your searching would be much more precise, and you'd get to the best answer more quickly.

Now I'm curious. What's the name of that muscle that's just to the outside of your shin bone? Can you figure out what kind of exercise might cause injury to that particular muscle?

6 What's the Most Likely Way You'll Die? How to Be Explicit about What You're Searching to Find (and Why That Matters)

Looking up data sources is fairly easy, but understanding how much you should trust that data is a bit trickier.

I don't mean to be macabre, especially after the previous chapter, but there are times when your thoughts turn to the eternal. The life of a soldier in the War of 1812 could end in a fairly dramatic way. What about we ordinary people? How do we live, and how do we die?

I was running through yet another cemetery, this time in Middletown, Connecticut, and reading the headstones made me wonder about what causes people die from each year. To make it personal, when I read the headstones in this cemetery, I started to wonder, what are the top reasons people die in the United States these days? We're not in the middle of the Civil War, so that's not what's causing deaths, but what are the leading causes today?

Before you answer that question, give it a thought for a second. What's your intuition about this? In the United States, what fraction of people die from car accidents? Is it as much as 10 percent of all deaths in a year? Or 15 percent? Or is it as low as 2 percent? How many people die from other kinds of accidents, like falls from a high ladder or slipping on a banana peel? Is that a significant fraction, or is it less than 1 percent?

What of different medical conditions? Do you have any idea what fraction of people die from heart attacks versus cancer versus infections? Which is a higher proportion of all deaths: medical causes or accidental causes? I realize that I don't know the answers to these questions, even though it's an important piece of data to know.

Figure 6.1
Cemeteries from the mid-nineteenth century show that people passed away from a surprising number of causes that we no longer think much about in the United States. The Civil War was one such cause, which we no longer worry about, along with malaria, yellow fever, and the flux.
Credit: Daniel M. Russell

Let's reframe my wondering with two specific research questions.

Research Question 1: *How many people die (from all causes) each year in the United States?*

Research Question 2: *What are the top five causes of death in the United States (as a fraction of the whole)?*

There are two things that I've learned about intuitions over all my years of studying human behavior. First, even though we love to think of them as highly accurate and useful, they're often wrong (especially about areas in which you have little expertise). And second, if you don't write your intuitions down, you'll quickly and unconsciously adjust them to align with the facts, once you figure out what they are. It's almost as though we can't stand to think of ourselves as making a mistake, so your mind will quickly back up and retroactively adjust your intuition to what you've learned. ("Oh … I *knew that.*" No you didn't. You just changed your mind once you found out the reality.)

Knowing this is true, before I did any research on these questions (and at the risk of exposing my complete ignorance about such things), I wrote down a few of my intuitions, amplifying the research questions just a bit to make them measurable and easy to test. This way, I'll know exactly what I learned.

Here's how I amplified the questions.

A. What fraction of people die from car accidents in the United States each year?

B. How many people die from other kinds of accidents?

C. How many people die of different medical conditions?

D. What are the leading causes of death?

I noted my guesses before having done any research.

A. Car accidents: I guess it's around 15 percent of all the US deaths per year

B. Other accidents not from cars: 5 percent per year

C. Medical conditions (not including old age): 50 percent

D. Leading causes of death (of any or all causes), in order: accidents, heart problems, and cancer

The obvious queries on different search platforms give different numbers. There's variation in the answers even within a single search platform. Compare these results with slightly different queries on Google.

Figure 6.2
Credit: Google and the Google logo are registered trademarks of Google Inc., used with permission

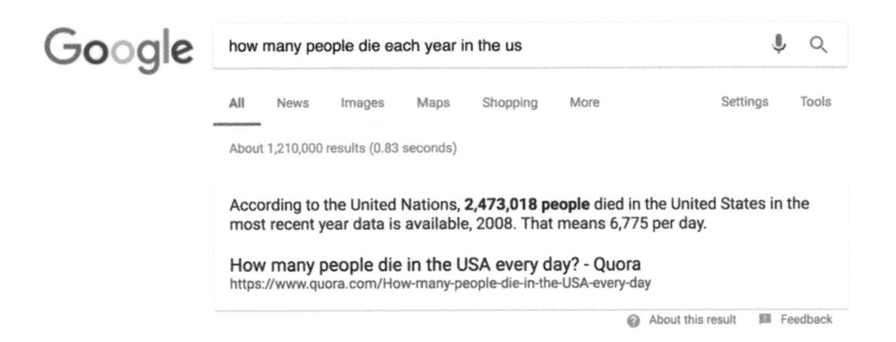

Figure 6.3
Credit: Google and the Google logo are registered trademarks of Google Inc., used with permission

Notice that there's a 250,000-person difference between these two numbers. Why? Because they come from different sources. The first query gives a short web answer from www.medicalnewstoday.com (which in turn gets its data from the 2014 Centers for Disease Control—aka the CDC—numbers), while the second query shows an answer that's from Quora.com with data from the UN data source UNstats.un.org, and these numbers are from 2008.

Oddly, the first article tells us that the CDC data is no longer available. The link that the *Medical News Today* cites IS broken, but there is an obvious query you can do:

[CDC 2014 data deaths]

It takes you to the CDC's "National Vital Statistics Report," which has exactly the same number: 2,626,418 in 2014.[1]

If you click on the Quora link in the second query [how many people die each year in the us], the write-up there takes you to the UN Demographics Report from October 2017, which tells us the total number of deaths for 2015.[2]

Looking at that page, you see the entry for the United States.[3]

READ CAREFULLY: The Quora article says that "the most recent data available is from 2008." But this data is from 2015 (the date is in the gray column), and the report was updated on October 16, 2017, yet notice that the number shown here is different from what's in the summary! Here, the United Nations says it's 2,712,630 deaths in 2015, as opposed to the

3. Live births, deaths, and infant deaths, latest available year (2002 - 2016)

Last updated 16 October 2017

Continent and country or area	Live births				Deaths				Infant deaths			
	Year	Code[a]	Number	Rate	Year	Code[a]	Number	Rate	Year	Code[a]	Number	Rate
AMERICA, NORTH (cont.)												
United States of America	2015	C	3 978 497	12.4	2015	C	2 712 630	8.5	2015	C	23 455	5.9
United States Virgin Islands[41]	2012	C	1 415	13.4	2012	C	723	6.9	2012	C	13	...

Figure 6.4
From UN demographics report.
Credit: United Nations, Department of Economic and Social Affairs Statistics Division

2,473,018 deaths reported in the 2008 UN summary seen in the web answer. Notice that we're comparing deaths in 2008 versus 2015; of course there's a big difference.

Think about what this means: certainly you'd expect the total number of deaths to change year by year; the overall population increases year by year, and the death rate changes as well—just much less than the overall growth in population.

OK, so now we have a slightly different question to answer: Can we find the CDC data from 2015 to be comparable with the UN data?

I noticed that in the CDC report I found above, the actual text in the paper was this:

> In 2014, a total of 2,626,418 resident deaths were registered in the United States.

I know that these kinds of reports are often written from a template. (That is, the CDC probably just copied the report and plugged in the new numbers for 2015.) So I did this query to find the report for 2015:

["In 2015, a total of * resident deaths"]

Notice that I changed the year to 2015 and used the * operator (for more details, see **how to use the * operator**[A]) to match the new number for that year, and I double quoted the whole thing to find a match for this exact phrase.

Voilà! That takes me directly to the 2015 CDC report, where we find out that "a total of 2,712,630 resident deaths were registered in the United States in 2015."

Let's compare these numbers from the United Nations and CDC:

2014

United Nations	2,626,418
CDC	2,626,418

2015

United Nations	2,712,630
CDC	2,712,630

Notice anything odd about these numbers? Both the United Nations and CDC numbers are exactly the same! If you go back a few years, you'll see more of this pattern. Which makes me wonder, Where does the United Nations get its numbers? From the CDC! (After looking around, I found that nugget in a footnote, of course.)

Which means that although you might think you've "double sourced" this data, this actually is NOT two different sources; the United Nations is just taking whatever data the CDC hands it.

You might be tempted to think that the United Nations is getting its data from a different US source; after all, it gives its data citation as coming from the "U.S. National Center for Health Statistics" in its "National Vital Statistics Report." But when you look up the National Center for Health Statistics, you discover that it's a department of the CDC. It turns out that it consists of the same people who collect the data in the CDC!

This is an interesting insight: the simple question, *How many people die each year in the United States?* turns out to have a more complicated answer. It varies by year, and as you might imagine, it varies depending on how you measure it.

WHAT? Isn't a death a death? Can't you just count death certificates?

Well, yes, but are you also counting people who disappear? What about US citizens who die overseas? Are they listed as a US death or a death in that country? Are you counting from January to January, or just one month-long period and multiplying by twelve? Are abortions counted as deaths? Are stillbirths counted? What about people in Puerto Rico, the US Virgin Islands, and other territories? (Why are the Virgin Islands broken out into a separate line item in the CDC report?) What about military deaths in non-US locations?

As frequently happens, once you start digging into a research question, you learn a lot about the area. You learn the little details about your question that deepen your understanding of the question you're asking. This happens all the time when we do our research questions; what starts out as a simple, straightforward question turns into something larger and with more nuance than you thought at the start.

In each of the questions I asked above, when looking for the definitions, you can often find the answers in the data commentary that's usually at the bottom of the data set. (Sometimes it's scattered around in the text itself.) But it looks like this, typically presented as footnotes in the article. Here's an example from that UN data set (figure 6.5).

35	Based on the results of the 2010 Population Census.
36	Excluding U.S. Armed Forces overseas and civilian U.S. citizens whose usual place of residence is c
37	Source: U.S. National Center for Health Statistics, National Vital Statistics Reports (NVSR).
38	Data refer to projections based on the 2010 Population and Housing Census.

Figure 6.5
Credit: Wikipedia

The notes describe the properties of the data; in this case, footnote 36 in that report tells us that military and US civilians who die outside the country are NOT included in the totals.

In this instance, we found out that which year you're asking about makes a big difference.

Now, what about that other question—the causes of death in the United States?

Those same reports also break down the causes by the percentages of all US deaths. From the CDC report on health issued in 2017 (with data from 2015), we find that the top five causes of death in the United States are:

1. Heart disease (23.4 percent)
2. Cancer (22 percent)
3. Chronic lower respiratory disease (CLRD) (5.7 percent)
4. Accidents (5.4 percent)
5. Strokes (5.2 percent)

The CDC illustrates this nicely with a chart (from the previous CDC reference). (See figure 6.6.)

As you can see, heart disease and cancer are the two largest causes of death, accounting for 45 percent of all deaths in 2015. CLRD, the next most common cause, is only around one-fourth as much.[4]

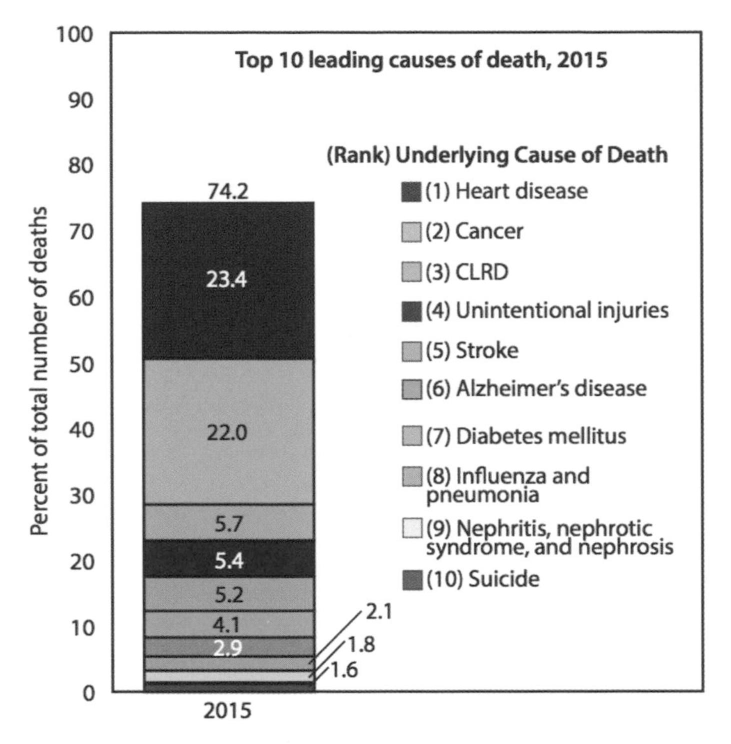

Figure 6.6
From a CDC report, National Center for Health Statistics, *Health, United States, 2016: With Chartbook on Long-Term Trends in Health* (Hyattsville, MD: National Center for Health Statistics, 2017), 18.
Credit: Centers for Disease Control

When I look back at my guesses (at the top of this post), I see my intuition was *really* wrong. Accidents of all kinds are around 5.4 percent of the total (which means that car accidents are less than that).

We may worry about mass murders or the latest version of the flu, but the big killers each year are heart disease and cancer. They are much more significant in terms of public health than anything else by far.

When you look at the causes of death over time from this same report, it's a fascinating piece of data (figure 6.7).

What is so striking is how constant many of these numbers of deaths are. Why do roughly the same number of people die each year in accidents?

This chart also has good and bad news: we're getting better at managing heart disease, but the overall cancer rate hasn't changed much in forty years.

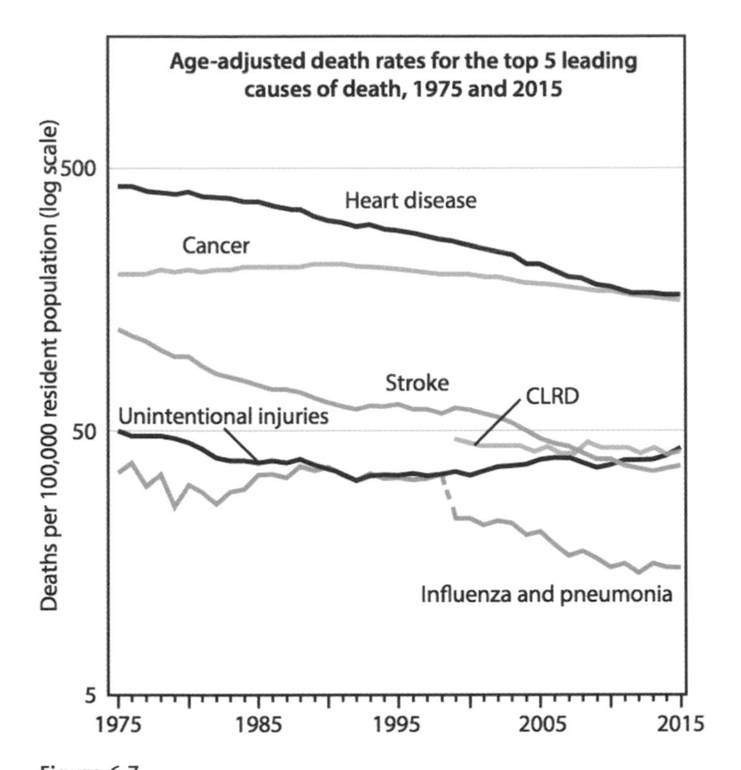

Figure 6.7

From a CDC report, National Center for Health Statistics, *Health, United States, 2016: With Chartbook on Long-Term Trends in Health* (Hyattsville, MD: National Center for Health Statistics, 2017), 18. Notice that the y-axis is a log scale, which means that a little bit of change going down (e.g., heart disease or stroke) is actually MUCH bigger than it might seem. That decline seems much less impressive than it really is. The improvement over forty years is amazingly good. Note also that CLRD is a new disease label that combines asthma, bronchitis, and emphysema. In 1999, the disease coding system changed to recognize those diseases as a cause of death, and separated out pneumonia and the flu into another category.

Credit: Centers for Disease Control

And of course, another big factor in the causes of death is age at time of death. People die of different causes at different ages. I saw a data table that suggested this, so I wanted to follow up and see if I could find a summary chart from the CDC. But how can I search *just* the CDC website? That's easy; you **use the site: operator**[B], as in this query to search only the CDC.gov site:

[site:cdc.gov causes of death by age]

That led me to this chart (table 6.1) in the CDC chart *collection* for causes of death in the United States, which shows how people die for different reasons at different ages.[5] While cancer and heart disease are the largest causes of death, they come into play only after age forty-four. Before that, you're more likely to die of an accident.

Research Lessons

1. *When looking at data, be SURE you understand WHEN it was collected and WHAT it's measuring.* As we saw, different sources all draw on slightly different resources from different times. This can make a big difference in your results. Remember also that if two sites have exactly the same data, they're probably both using data from the same source, and that's not the same as getting two different perspectives on the data.

2. *Consider other factors that might influence your data; in this case, death rates vary a LOT by age.* (They vary by other factors too, such as gender, race, and location—but I just focused on age here.) Be sure you understand all the aspects of the data that are important to you.

3. *When you need the "next document in the series," remember that those documents often use boilerplate language, which you can find with a fill-in-the-blank query like* ["In 2015, a total of * resident deaths"]. This is an amazingly handy trick to remember when you're looking for documents that have patterns of text that you know. You can use the * to fill in the missing text as a way to search for that kind of content, which is especially useful for series.

4. *Be sure you know where your data comes from!* I naively thought that the United Nations would have different data than the CDC, but noticing that its numbers are all the same drove me to check where the UN data came from—and it was from the CDC. This data is NOT truly double sourced!

Table 6.1

Top Five Causes of Death by Age Group 2015

Rank	< 1	1–4	5–9	10–14	15–24	25–34	35–44	45–54	55–64	65+
1	Congenital abnormality: 4,925	Unintentional injury: 1,235	Unintentional injury: 755	Unintentional injury: 763	Unintentional injury: 12,514	Unintentional injury: 19,795	Unintentional injury: 17,818	Malignant neoplasm: 43,054	Malignant neoplasm: 116,122	Heart disease: 507,139
2	Short gestation: 4,084	Congenital abnormality: 435	Malignant neoplasm: 437	Malignant neoplasm: 428	Suicide: 5,491	Suicide: 6,947	Malignant neoplasm: 10,909	Heart disease: 34,248	Heart disease: 76,872	Malignant neoplasm: 419,389
3	SIDS: 1,568	Homicide: 369	Congenital anomality: 181	Suicide: 409	Homicide: 4,733	Homicide: 4,863	Heart disease: 10,387	Unintentional injury: 21,499	Unintentional injury: 19,388	Chronic low respiratory disease: 131,804
4	Maternal pregnancy complication: 1,522	Malignant neoplasm: 354	Homicide: 140	Homicide: 158	Malignant neoplasm: 1,469	Malignant neoplasm: 3,704	Suicide: 6,936	Liver disease: 8.874	Chronic low respiratory disease: 17, 457	Cerebro-vascular: 120,156
5	Unintentional injury: 1,291	Heart disease: 147	Heart disease: 147	Congenital anomality: 156	Heart disease: 997	Heart disease: 3,522	Homicide: 2,895	Suicide: 8,751	Diabetes mellitus: 14,166	Alzheimer's disease: 109,495

Credit: National Center for Health Statistics, *Health, United States, 2016: With Chartbook on Long-Term Trends in Health* (Hyattsville, MD: National Center for Health Statistics, 2017).

5. *Check your intuitions at the door when you do research.* My intuitions about causes of death were embarrassingly bad. I wrote them down at the start of my research to make me understand what it was I was learning, and see the distance between where I started with my research and where I ended up. Although this is painful, seeing your mistaken intuitions in black and white is a strikingly powerful learning moment.

How to Do It

A. How to use the * operator. In my above example, I used the query:

["In 2015, a total of * resident deaths"]

Here the * is used as a kind of fill-in-the-blank operator. For this query, we're looking for a number that fills in for the * in the query.

This is a handy Google search trick when you're not sure of a word (or two) in a phrase that you'd like to find. Another use would be to find organizations when you can't remember their name. For example, you might search for an organization that's at a university, but is the Department of X Technology. You know you'll recognize it, but you just can't come up with the middle words. Use the * operator in a query like:

["Department of * Technology" University]

Such a fill-in-the-blank query gives you the chance to see what kinds of possibilities there are.

B. Use the site: operator. Truthfully, I probably use site: more than any other operator in my searches. It just restricts the search to a particular site, which is often exactly what I want to do. Above I used the query:

[site:CDC.GOV causes of death by age]

It allowed me to search for results with the words "causes of death by age" ONLY from the CDC's CDC.GOV site. You can limit your searches to a particular online location (like site:CDC.GOV or site:Stanford.edu) or even particular countries with a query that specifies the country code (like [influenza site:UK] to search only sites within the United Kingdom.

Try This Yourself

It's relatively easy to find data, but it's sometimes hard to make sure you've got the *right* data. In particular, if you're ever going to read the fine print on something, be sure to read the fine print on the metadata (that is, the information that describes the data set) before you do anything with it. Even simple research questions can contain surprising depths.

For instance, how would you find out the answer to the question, How many different breeds of dogs (or cats, depending on your personal preference) are there?

Knowing how to use the **site:** operator, you can look for the answer to that question by looking inside specific sites that you trust. When you do this search, read the details of the definitions carefully. What is the definition of a "breed"? (Is a cockadoodle a breed of dog? Is the Savannah cat really a domesticated cat?")

As you search for the answers, think about why you trust a particular organization's site. Is the American Kennel Club, for example, an organization that has a good definition of *dog breed*, and if so, why do you think so? Would you believe the German Kennel Club (VDH.de) more or less than the American Kennel Club (AKC.org)? So, how many breeds of dog are there? Do you trust the source you find? Why?

7 When Would You Want to Read the Italian Wikipedia? How to Look for Information from Other Languages in Wikipedia and Other Sources

Sometimes, looking for information in another language than yours can be immensely useful.

I was talking with a friend about the stories of Leonardo da Vinci, that great polymath of the Italian Renaissance. It seems there's some debate about whether he was the greatest artist of all time, or merely a worthy competitor to Michelangelo and other contemporaries.

As we talked, the legacy discussion became a bit heated. How could one get a sense for the man some five hundred years later? Naturally, my thoughts turned to how I would do this piece of research online. It's easy enough to find all kinds of online resources, but is there a way to get a sense of the legacy of a famous person? What would that even mean? I decided to capture my curiosity and write down my research question.

Research Question: *How can you get a good sense for different cultural interpretations of an idea (in particular, for Leonardo and Michelangelo)? Is there a way to do this with online research?*

This is a fairly open-ended question, and it's the kind of thing you can spend a long time answering. A historian might say that you need to read about his life and works extensively. But that kind of time to devote to a single issue is a bit of a luxury. Is there something else one could do? What about Wikipedia, that grand online encyclopedia? How could we use it effectively?

Figure 7.1
Often, the fastest way to search Wikipedia is to do a Google search with the context term "Wikipedia" as part of the query. This not only gives a wide range of results from Wikipedia but also a set of deep links into the Michelangelo article.
Credit: Daniel M. Russell

There's a lot of debate about how accurate and reliable Wikipedia is, but the bottom line for me is that for many topics, it's fairly reliable.[1] This is especially true once you know a bit about *how* to do online research with Wikipedia.

If you're not already in Wikipedia, the fastest way to search Wikipedia is to do a query like this, adding the context term "Wikipedia" to your search query. The results will be primarily from Wikipedia and Wikimedia.

If the topic that you're searching on is fairly rich (as this one is), then you'll see an indented set of items. These are "deep links" into a specific section of the article on Michelangelo; click on them to jump into that section of the Wikipedia page.

If you want to search ONLY within Wikipedia, you can use the **site:** operator (see chapter 3, where I show how to do this) to search exclusively inside Wikipedia. The results will be limited only to Wikipedia articles.

Now to find out about the possible relationship (or lack thereof) between the two men, you can list both of their names in the query. Scanning the snippets below each result gives a quick overview of their relationship (but don't stop reading there!).

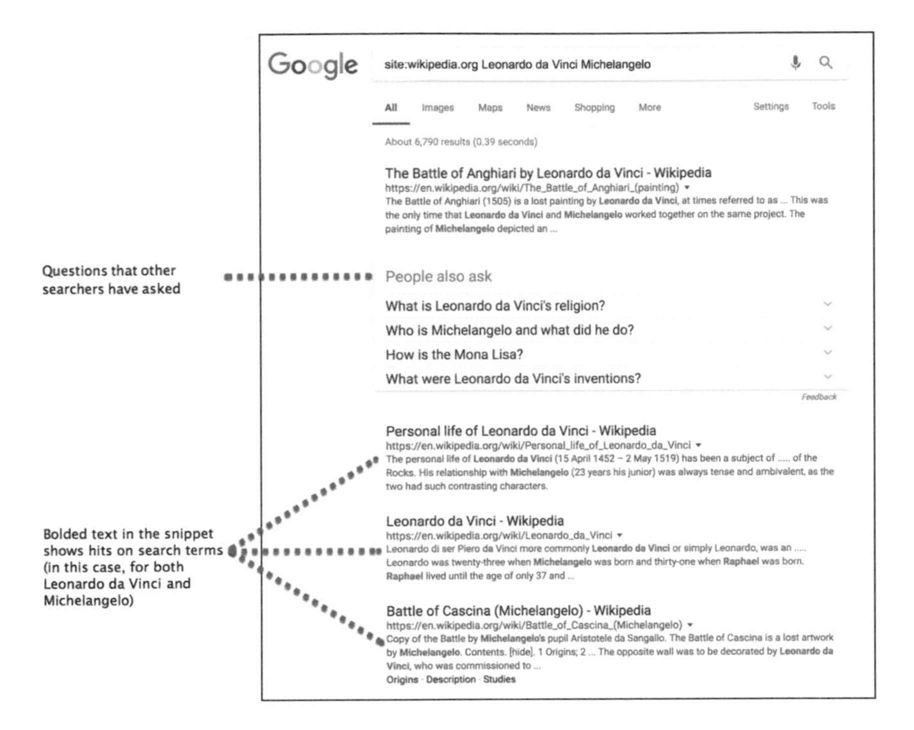

Figure 7.2
Use a **site:** operator to search only in the site, and use both names to find articles that mention both of them.
Credit: Google and the Google logo are registered trademarks of Google Inc., used with permission; Daniel M. Russell

Doing the search [site:Wikipedia.org Leonardo da Vinci Michelangelo] *with* the two artists' names will find articles where both are mentioned within Wikipedia. It's also useful, perhaps as a second step, to do a search for both names without the Wikipedia **site:** restriction; that will find many articles about their relationship. Starting with just Wikipedia, though, is a great way to quickly learn about their connections.

From the articles on this page, it's easy to quickly learn that the two men were both highly regarded during their time as strikingly talented artists. What surprised me, however, was learning about the painting of the *Battle of Anghiari* (by Leonardo) and *Battle of Cascina* (by Michelangelo)—the only instance when two paintings by both masters were painted at the same time.[2] Neither painting was finished, and Leonardo's seems to be buried

Figure 7.3
Searching for the relationship between Michelangelo and da Vinci is another way to get an overview of their interactions. (Of course, be careful about what you read; the first hit here is a novel about the two men. While entertaining, it might not be what you want.)
Credit: Google and the Google logo are registered trademarks of Google Inc., used with permission

beneath later work. But writers at the time judged both incomplete works as evidence of great masters.

And while usually named together as the two giants of Renaissance art, Leonardo and Michelangelo just barely worked at the same time. Leonardo was twenty-three when Michelangelo was born and was artistically active for thirty-seven years. Michelangelo began his professional career when he was seventeen and was creative for around fifty-two years, although often in sculpture and architecture rather than painting.

In the same way, you could do a query on Google Books to search for more articles in depth on this topic. Here I've done a search on Google Books for the relationship between Michelangelo and da Vinci (figure 7.3).

This is a bit like the **site:** search in that when you search Google Books, the query will only return books that have a hit.

I could go on in more detail about who these Renaissance giants were and why they seemed to have a particularly awkward relationship, but my point here is really to show you how to use online research tools to understand the relationship between two people (famous artists in this case). Wikipedia is a great place to start your research, especially when you realize that Wikipedia has more depth than you might have imagined. In particular, there are a few things worth knowing about how Wikipedia is organized.

Why You Might Want to Read the Italian Wikipedia

You might have been using Wikipedia for quite a while and not noticed that there are entire other worlds of information there on the page for you to examine. Take a look at the left side of this Wikipedia entry about cats. Do you see the column over on the left labeled "Languages" (figure 7.4)?

Wikipedia comes in 287 languages (and more are being added every day). Those links on the left point to the equivalent article in another language.

Notice that the gold star next to the language means that this is a "featured article," which are given a star by the editors, indicating that this article is of exceptionally high quality and often used as an example for other Wikipedia authors.[3] It's thus frequently worth checking out articles in languages with a star.

What you might not realize is that articles are NOT THE SAME in all the world's different languages. Even fairly straightforward articles, like this one on "Cat," can be different, and as such, well worth looking at for comparison purposes.

Here, for example, are the outlines of the cat article in English (on the left) and the Spanish Wikipedia article about cats (on the right, in translation). Note how different they are (figure 7.5).

For some reason, the Spanish authors of the "Cat" entry go into MUCH greater detail about cat diseases (sections 9.1–9.13) than the English authors (where it's only covered in section 7.1). I have no idea why the English-language entry is so much shorter. Are Spanish-language cats more likely to get sick? That seems silly. But the Spanish-language edition covers, for instance, the effects of secondhand tobacco smoke on the rate of oral cancers among cats. (No surprise here: secondhand smoke is bad for your cat as well.)

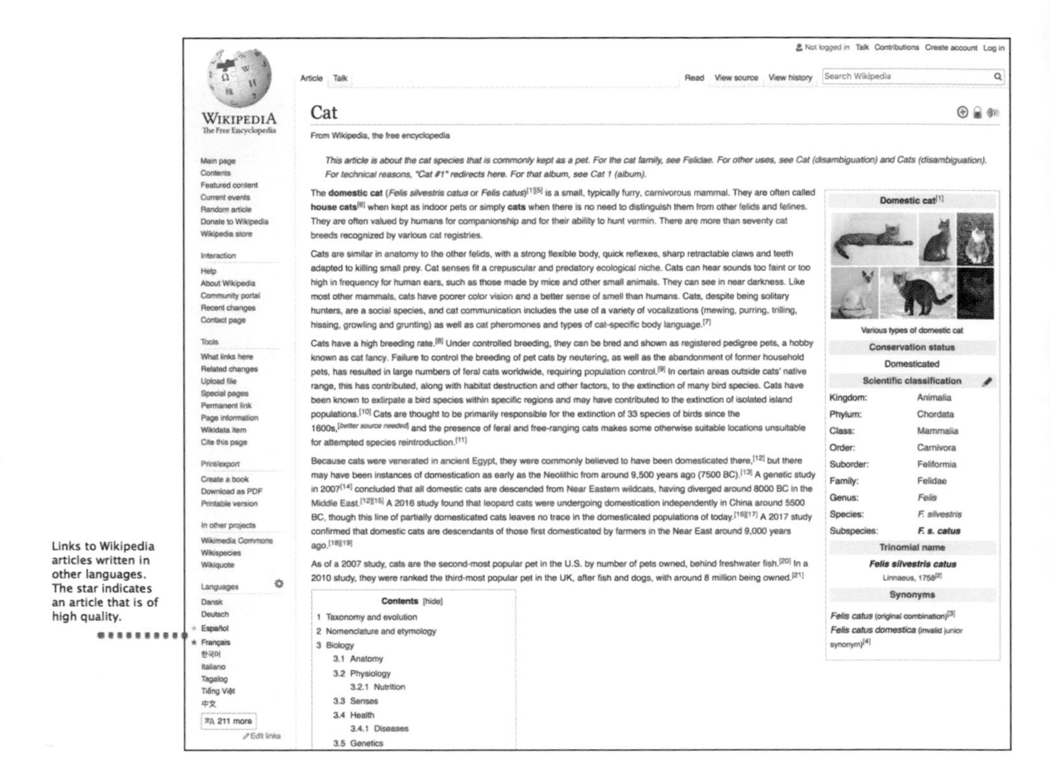

Figure 7.4
Wikipedia offers multiple versions of an article in different languages. Here you can see many languages that have an article about cats. Notice the stars next to some languages; those are high-quality articles. It's often worth checking out articles in languages with a star as they're usually quite good.
Credit: Wikipedia

Bear in mind that the topic of "cat" is fairly noncontroversial. When you compare more culturally, historically, or politically loaded topics, the differences become even greater. Consider for a moment how controversial topics are handled: the Wikipedia editorial board locks down articles that tend to attract the most vandalism. (The article about Hilary Clinton is one such example. Ordinary people can't edit it; only fairly trusted Wikipedian editors can.) But what's more, all these different languages can give you, the online researcher, a set of different perspectives on a topic.

Now we can get back to my original topic of comparing Leonardo and Michelangelo.

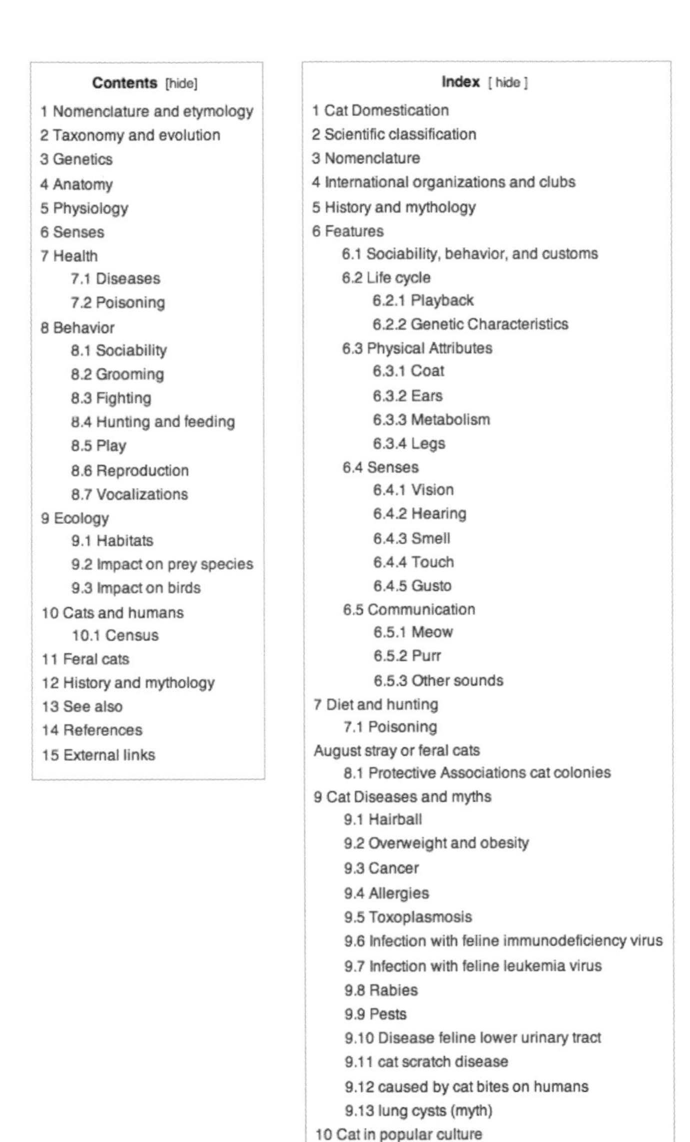

Figure 7.5
Wikipedia articles on "Cat"; the English version is on the left, and the Spanish version is on the right
Credit: Wikipedia

Figure 7.6 is a side-by-side comparison outlining the English and Italian versions of the Wikipedia articles about "Leonardo da Vinci." It makes sense that the Italian version would be much more comprehensive, but it's also interesting to read the rather-different treatments that each culture makes about his personal life and relationships.

Notice that just section 1 on the life of Leonardo in Italian (twelve-thousand-plus words) is longer than the entire article about Leonardo in English (a mere eight-thousand-plus words).

And this difference becomes especially obvious when you look at the Italian version of the article about Michelangelo's *Battle of Cascina*; it's a highly detailed and well-documented article. The corresponding English article about the painting is barely three hundred words.[4] These kinds of cross-cultural comparisons are fascinating to make, but also potentially really valuable when doing your research. If you're frustrated by the lack of depth of the Wikipedia article in your own language, try another language, perhaps one that's "closer to home" for the material. More important, you'll be getting different views on the topic.

This works for people, institutions, places, and cultures, yet it's oddly the opposite for some deeply cultural concepts.

For instance, suppose you want to quickly get an understanding of the Danish idea of *hygge* (which translates into English, roughly, as a "mood of coziness and comfortable conviviality with an overall feeling of wellness, quiet happiness, and contentment"). Looking up hygge in the English Wikipedia gives you a pretty good idea of what hygge means, but checking out the Danish version is a disappointment. It's simple enough to get the translation into English, but the Danish article is fairly short and feels like the sketch of hygge. By contrast, both the English and Italian hygge articles are much longer and more complete (with the Italian one being especially detailed, including a discussion of hygge in nondomestic settings and even recipes for food that are particularly hygge-ish).

When I'm looking to understand a person, place, or thing that is particularly embedded within a culture (or language), then the Wikipedia articles in that language tend to be good. But when looking up a less tangible concept—like Danish hygge, gemütlichkeit (the German version of coziness), or *jugaad* (the Hindi idea of a quick fix or improvised repair)—these tend to be more expansively defined in any language *other* than that of its origin. The English Wikipedia article on jugaad is quite good, as is

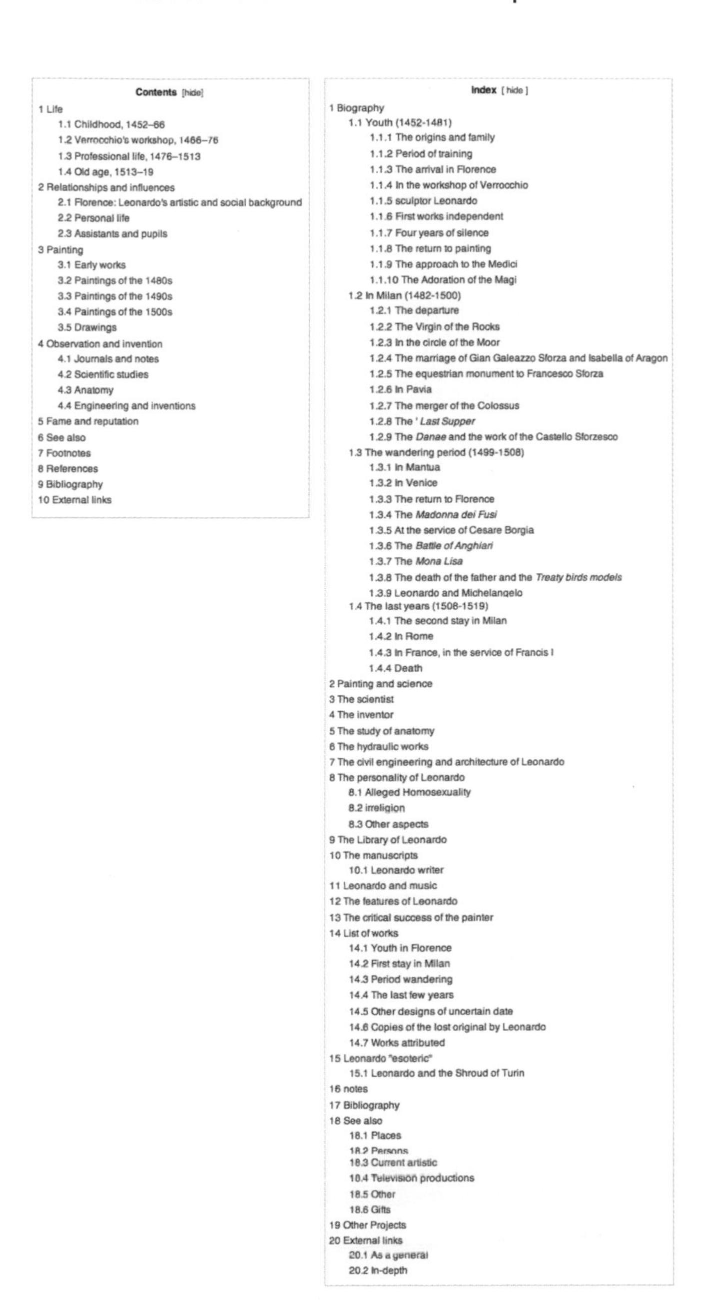

Figure 7.6

Comparing the outlines for both the English- and Italian-language versions of Wikipedia articles on Leonardo da Vinci.

Credit: Wikipedia

the French version of gemütlichkeit, but if you read the German article on gemütlichkeit, you'll probably be disappointed.

A Tool for Comparison

A rule of thumb that I use when I find myself trying to do something repeatedly is to *look for a tool to help with the task.*

In this case, after noticing that I was bringing up multiple side-by-side browser windows to compare Wikipedia articles in different languages, I thought about my rule of thumb. How could I find an online tool that would help me compare two articles side by side? The first query I could think of was:

[tool to compare Wikipedia articles]

This leads, naturally enough, to several hits to pages on Wikipedia that list the features of different file comparison tools; that's nice, but not quite what I want.

But a bit further down the page I found the Manypedia.com site, which does exactly what I wanted. It lets you look up a single topic and then look at the Wikipedia articles from different languages side by side. For example, if we look up a topic we're interested in, say, the Italian sculptor Gian Bernini (who was born just seventy-nine years after Leonardo died), we can see the English and Italian (in translation) versions side by side (figure 7.7).

I've added two numbers to the screen capture to point out a couple features in the Manypedia interface:

(1) is the list of all the images from the Wikipedia article (each shown as a small thumbnail image). Notice how the list of pictures is VERY different between the two articles. His great work *The Hermaphrodite* isn't even mentioned in the English version, yet it shows up in the Italian article. You can spot this immediately by looking at the images summary and noting the differences between the sets of images.

(2) is a kind of word cloud (the font size indicates how often a word appears in the article) just below the images summary for the Wikipedia entries. Again, a quick scan shows real differences between the articles. In the Italian article, "Lorenzo" (which appears 47 times) and "Barberini" (which shows

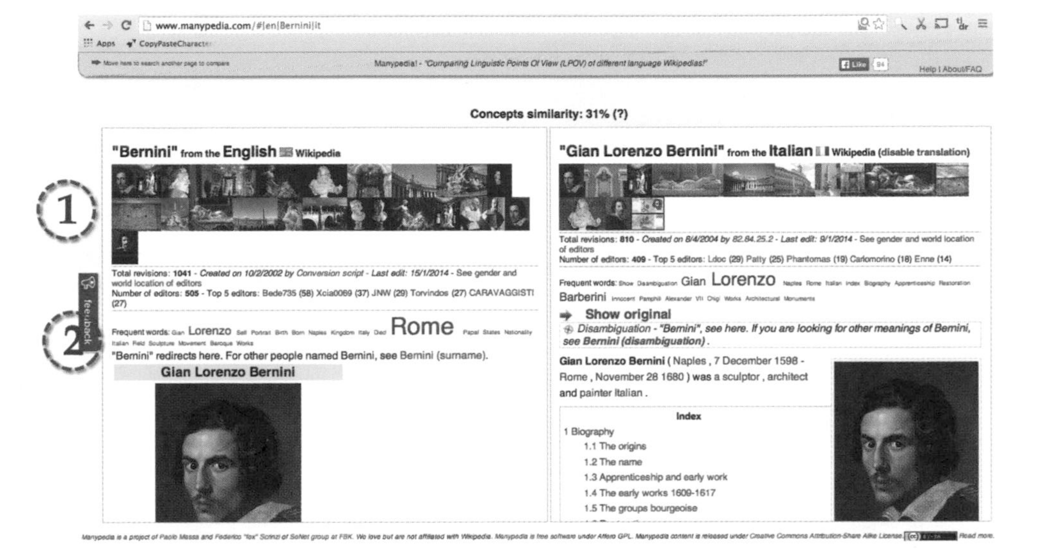

Figure 7.7
With the Manypedia web app, you can compare different versions of the same Wikipedia entries from different languages. The side-by-side comparison highlights the different values that each culture places into its Wikipedia entries.
Credit: Manypedia

up 14 times) are more important, but by contrast in the English version, "Rome" (appearing 106 times) and "Lorenzo" (found 47 times) dominate.

If Manypedia shows context between two different language versions of the same page, Wikipedia itself has another way to show context across ALL Wikipedia.

What Links Here

Wikipedia also has a clever method for showing a bit of context that goes beyond the ordinary. The Wikipedia tool "What links here" appears in the left-hand list of links on an ordinary Wikipedia page (figure 7.8).

When you click on "What links here," you'll get a long list of all the other Wikipedia pages that have a link to this one—in this case, all the other Wikipedia pages that link to Leonardo's page.[5]

Why should you care? Because scanning that list is effectively an overview of the other people, places, times, projects, and influences that

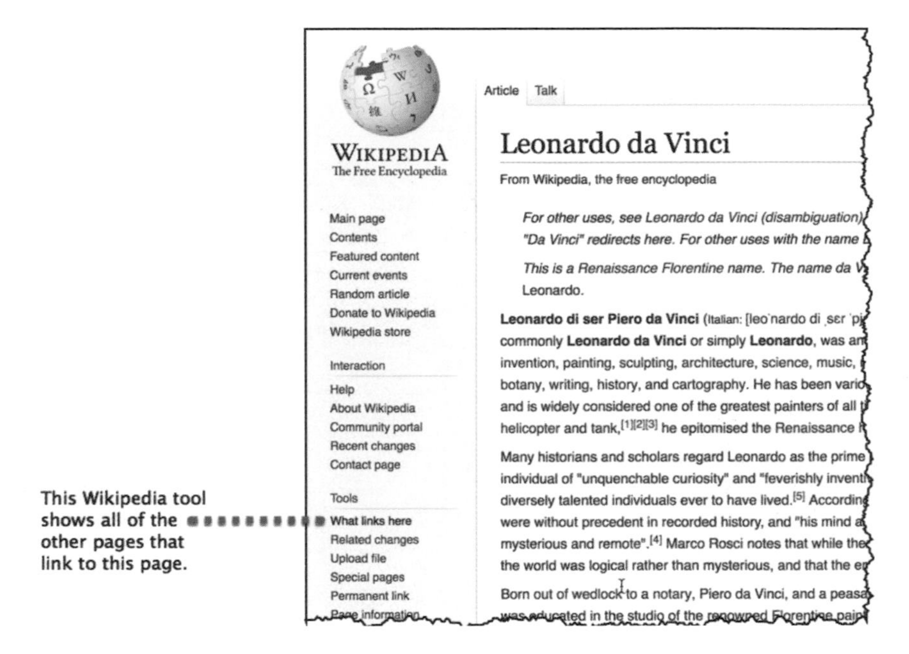

Figure 7.8
Several Wikipedia tools appear in the left-hand column, including this one that opens up a new page with a list of all the other Wikipedia articles that are linked to this article.
Credit: Wikipedia

somehow connect to Leonardo. It's as though someone pulled out all the key concepts from Leonardo's article and presented them to you as a list. I often look at this list when I'm trying to get a sense of what connections there are between my main topic and the rest of the world (figure 7.9).

For instance, reading through the list of pages that link to "Leonardo da Vinci," I can see that there's some connection to Cesare Borgia, a well-known person of the age. By rolling over the link, a pop-up summary appears, and if it looks interesting enough, you can open this article in a new browser tab.

Research Lessons

There are a few lessons here about using Wikipedia as a research tool.

1. *Wikipedia comes in multiple languages.* When you're looking up an article on a topic that is highly culturally specific, be sure to look in that language

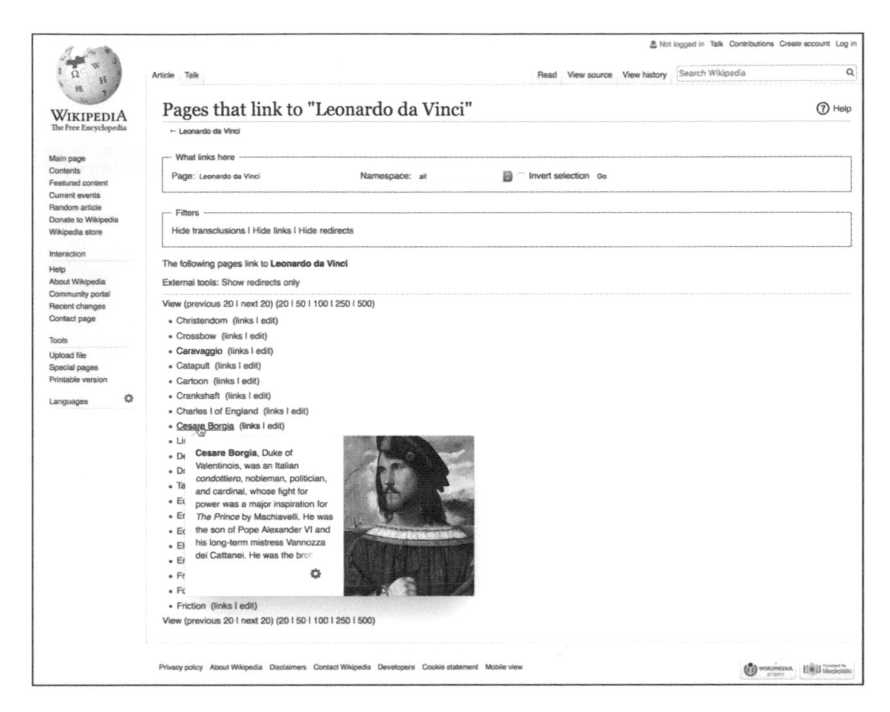

Figure 7.9
Hovering your mouse over the link will give you a pop-up preview of the linked page. Here you see a summary of Cesare Borgia, one of Leonardo's patrons in 1502 and 1503.
Credit: Wikipedia

as well. Famous people, places, and things tend to have longer and more detailed descriptions in the language that they're associated with. (Italian for Leonardo, for instance.)

2. *Pay attention to Wikipedia stars.* The stars mark articles that are especially well written and complete, and have good references. If you're looking at an article that seems a bit sketchy or lightweight, look for the same article in another language; if it's got a star, there's a good chance that reading the starred article (in translation) will be more useful than the inadequate article you're reading now.

3. *Search for a tool when you find yourself doing something repeatedly.* This is good advice in general; when you're doing something multiple times, this is a candidate for a little computational help. (As a simple example, if you're doing a lot of math over and over again, stop what you're doing and

learn how to use a spreadsheet.) In our case, there are online tools that can help compare multiple Wikipedia pages, making your learning process a bit more streamlined.

4. *"What links here" is a great way to scan the surrounding people, places, and things that have articles linking to your main topic.* The list of linked articles is sometimes surprising; it's worth checking out when you're trying to get a quick overview of a topic. At the very least, you'll find a list of the people and places that figure into the story.

Try This Yourself

Check out a few Wikipedia articles in languages that are appropriate for the topic, but in a language you don't read, write, or speak. You can use Google Translate to convert the articles into a language you DO read.

For example, earlier in this chapter I mentioned Danish hygge, gemütlichkeit (the German version of coziness), and jugaad (the Hindi idea of a quick fix or improvised repair). Look at the differences in each of these concepts in your own language (say, English) and the language of origin (Danish, German, or Hindi). What do these different Wikipedia articles tell you about the cultures of each idea?

Another comparison to make is between the English, Turkish, and German versions of the Wikipedia articles about *Göbekli Tepe*, a prehistoric site from around 9100 BCE found in central Turkey. Why are the articles in each language different? What do these differences tell you about the cultural perspectives of each? Can you find another culturally specific example of this cross-language effect?

8 Why Are the Coasts So Different? How to Use Online Maps Resources to Answer Broad Geographic Questions

The West and East Coasts of North America look quite different. Why is that? The west coastline has few islands just offshore, with lots of dramatic cliffs. The East Coast, by contrast, has lots of long sandy beaches with mellow islands and places to hang out. But why?

The other day I was standing in a friend's office at work. On his wall he had a large, beautiful, and high-resolution map of North America. Interestingly, it was black and white, which makes your eye see shapes of landforms in a different way. You don't get distracted by all the colors and annotations of the political map, or the colors of deserts, grasslands, and forests on a physical map. You just see the shape of the land itself; it's a kind of clarity of vision that removes all human traces, and you see just the shapes—the topography—pure and plain.

As I stood there looking at it, I noticed something I'd never seen before: that is, the East and West Coasts of the US are quite different. Yes, I know they're different culturally (Los Angeles versus New York, TMZ versus NPR, etc.). But what I saw for the first time was how *topographically* different the coasts are from each other. The eastern shores seem crinkly, full of long beaches and low-lying islands just off the coast. That surprised me. You think that I would have noticed before this.

Here are a few side-by-side images from Google Earth to give you a sense of what I mean. I've taken three overview shots at different resolutions so you can compare them. In each side by side, the scale is the same (figures 8.1, 8.2, and 8.3).

In this overview of both coasts, what I saw was that the East Coast has LOTS more islands and barrier reefs than the West Coast. That struck me as being really, really different.

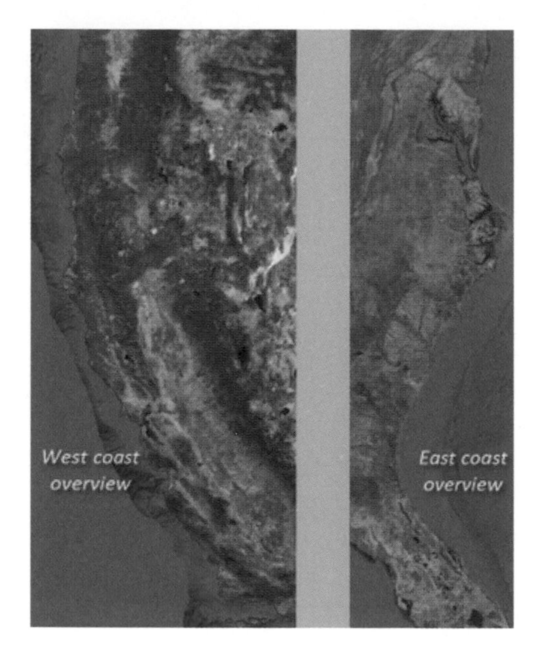

Figure 8.1
When you put the West Coast next to the East Coast, the topography looks quite different. The question is, Why?
Image © Google

Here's an even more zoomed-in view of two stretches of shoreline taken at the same latitude, east versus west. You can see that the eastern shore has many long barrier islands, with lagoons and waterways just west of the beaches (figure 8.2).

Do you see what I see? There are lots and lots of low-lying islands on the East Coast than on the west. (Note that I didn't try to cherry-pick really good locations; these samples are from the same latitude on the West and East Coasts, and are pretty representative of the topography.)

Finally, here's a real close-up of two locations, west versus east. Now the difference is really clear (figure 8.3).

Of course, there are long stretches of sandy beach in the west (think of the classic beaches of Los Angeles County), and stretches of stony beach in the east (think of the coast of Maine). But when you look at the long view (which is what that map made clear to me), you see different kinds of spaces. It's literally a different view.

Figure 8.2
Two different stretches of the west and east coastlines. Note how the west doesn't have many barrier islands, while they're common in the east.
Image © Google

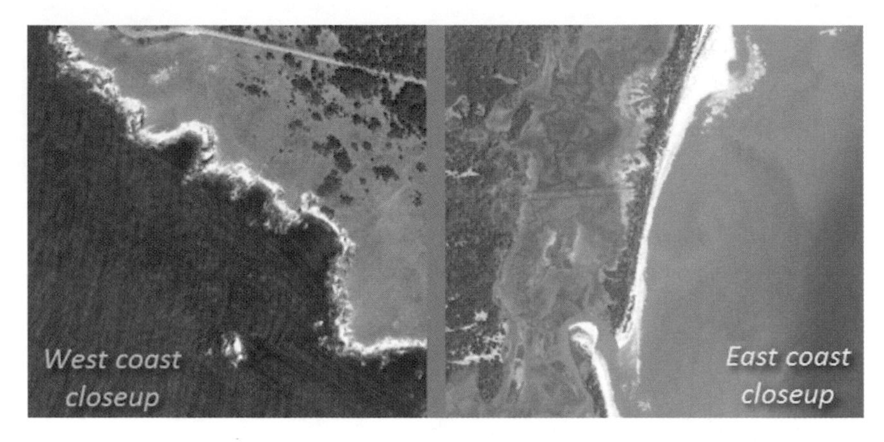

Figure 8.3
See that? The West Coast beaches are often stony and highly contoured. East Coast beaches are often broad sweeps of sand.
Image © Google

And now for the real question.

Research Question 1: *Why?*

Or to be more precise (because framing your research question carefully is half the battle of doing research):

Research Question 2: *Why is the East Coast topography dominated by low-lying fringe islands, while the West Coast of the United States is mostly without islands?*

To be sure, there are a few islands off the West Coast, but they're pretty jaggy and stony, while the East Coast islands are mostly sandy and flat. They are, in a word, "fringy," long and stringy in appearance, apparently mostly made of sand.

I started this research by trying to pick up a few terms that would be useful in my searching. I opened Google Maps and zoomed in for a while, looking at much of the East Coast to get a sense of what it's like. (This is something you can do easily with a zoomable user interface. It's simple to look at a whole continent, and pan up and down, while zooming in and out to get both a local and global view of the landforms.) I was looking for a typical section of the coast, and settled on a place on the coast of Maryland and Virginia.

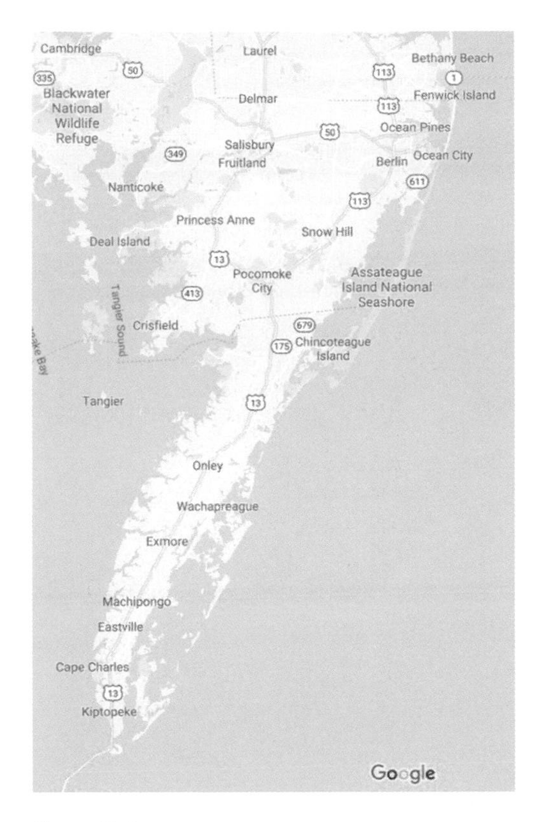

Figure 8.4
A typical stretch of the East Coast, with long, linear islands on the shoreline, lots of tiny islands, and water showing to the west of the islands.
Credit: Map data © 2018 Google

I chose Assateague Island as a representative land feature. Not only was it on a long and fairly consistent stretch of the East Coast, but I recognized the name from books I'd read long ago. And visually it seems like a representative stretch of the eastern shore. Since I knew there were a lot of books about Assateague and Chincoteague, I started with a query that would focus the results on what I wanted to learn about—that is, the geography:

[Assateague geography]

I quickly learned that these long, linear islands are called "barrier islands," and they're formed by sand being deposited along the shore by currents and storms that create a constantly changing arrangement of sand dunes, sandbars, and sand islands, and waterways between them all.

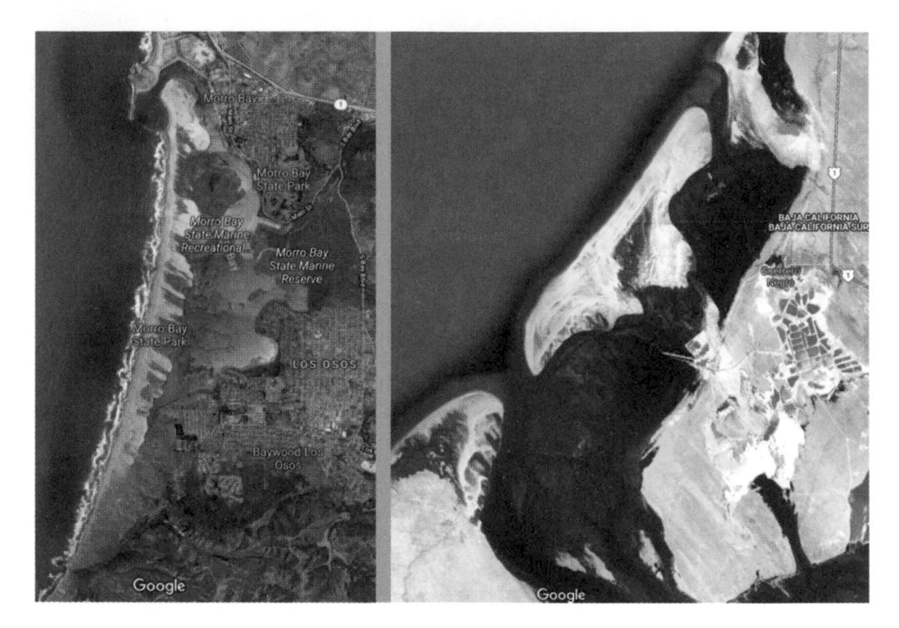

Figure 8.5
There are barrier islands on the West Coast, but there aren't many. On the left is Morro Bay in central California, and on the right is Guerrero Negro in Baja California. Credit: Google Earth

So now that I know what these long filaments of sand are called, I can reframe my question to ask a more carefully crafted version of my goal.

Research Question 3: *Why are there so many barrier islands on the East Coast, and so few on the West Coast?*

With online maps such as Google Maps that let you easily zoom in and out as well as roam up and down the coast, you can check for barrier islands just by zooming out and then dragging your view around. When you do this along the West Coast of North America, you'll find that there are a couple of barrier islands; there's one in Morro Bay in central California and another down south in Baja California in Guerrero Negro, but that's about it. If you look at the islands off the West Coast, they're all rocky points that seem like mountaintops rather than sandy bars.

The big surprise here is that out of all those roughly three thousand miles (or forty-eight hundred kilometers) of coastline, from Alaska to the tip of Baja California, you can scan the entire coastline using Google Maps and find just a few. By contrast, most of the East Coast is lined with barrier islands.

In my reading of these results, it became clear that the biggest difference is that the shorelines are different between the east and west. If you look at figure 8.1, you can see that the flat areas next to the coast are much wider on the East Coast (up to ten times wider) than the equivalent flat areas just offshore on the West Coast. Why? It turns out that the West Coast is the leading edge of the North American plate, while the East Coast is the trailing edge. Barrier islands form on the trailing edge because there's a broad, flat place for sand to accumulate.

The Pacific and North American plates meet on the West Coast at the *subduction zone*—that is, where one plate is sliding under the other. Any sediments that are carried to the ocean move across the narrow continental shelf (off California, this is about twenty miles wide) and fall into this area. There is no sediment buildup and consequently no barrier islands.

On the trailing edge (the East Coast), sediments carried to the ocean have a broad plain to accumulate on. New land is formed at the mid-ocean ridge as the plate moves and a wide continental shelf is created. The continental shelf off the coast of Georgia is around eighty miles wide, or four times larger than off California. And so barrier islands have space, a shallow continental shelf, and sand supply to form.[1]

So in the final analysis, here's the short answer to my original question, Why? Because plate tectonics (specifically, the nearby edge of the Pacific plate) force the western coast to be steep without much chance to accumulate the sediments needed to make barrier and fringing islands. By contrast, the East Coast has a long, gently sloping grade from the coastline to the next plate edge (which is far away), giving sediment an easy place to accumulate.

See the following map (from the Wikipedia article on plate tectonics; figure 8.6).[2]

Note that the Pacific plate butts up against California, while the North American plate extends way out into the Atlantic. That's a huge difference, and vastly changes the nature of the coastlines as well.

Research Lessons

The real lesson here is obvious.

1. *Look up stuff that's interesting, unknown, or unclear.* Learn to recognize the terms and concepts that you might not already know. In searching and

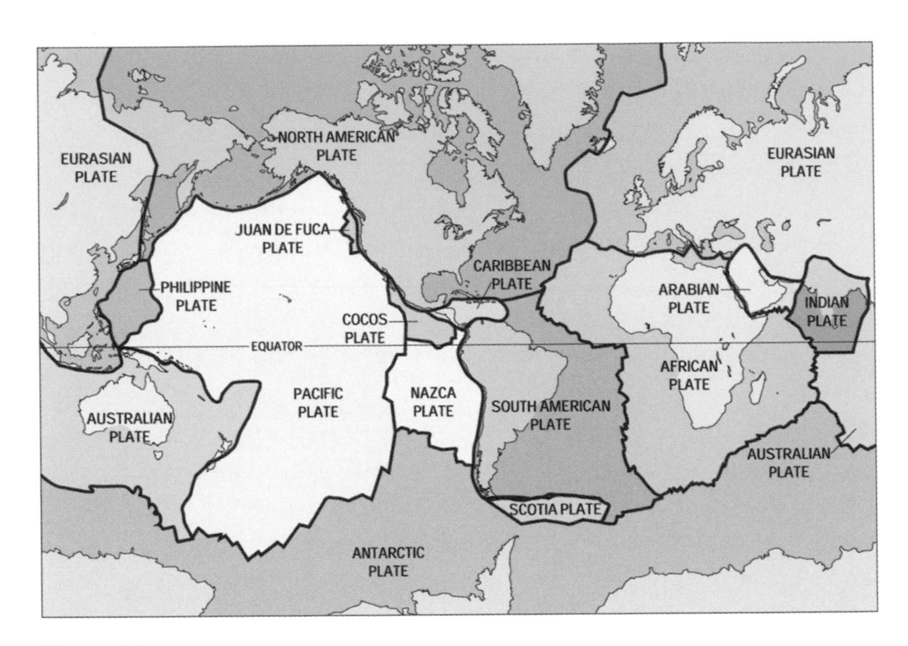

Figure 8.6
Credit: US Geological Survey

reading web pages to answer this question, I had to constantly look up words that were "terms of art." What actually IS a barrier island? What are "plate tectonics," and what is the difference between topography and geography? (You could just look them up—and you should.)

2. *You need to follow the search deeply enough to get a "real" answer.* Many researchers who work on this question found that the East Coast clearly accumulated sediment, but couldn't say why. Intriguingly, they said they could make up (that is, rationalize) a story quickly. But they didn't have any data to back up what they were saying. Word to the wise: when you don't have data, you usually know it. That's when you slip over the slippery slope into fake storytelling. It's easy to do, and we all practice it all the time. As researchers, though, we need to be aware of when this happens, and be able to stop ourselves and realize that this is the time for more research.

3. *Starting this search was hard; don't get discouraged at the start.* It was difficult getting started with searches that worked. In this case, unless you're a practicing geologist, terms like "continental shelf" and "shoreline sedimentary systems" don't just come trippingly off the tongue. Don't worry about

the terminology when beginning a research quest; eventually you'll learn the relevant language.

4. *Follow the people chain.* I did find several articles that helped me in my search. By starting with [west coast vs east coast barrier islands] I found an article comparing the two coasts by Molly Samuels (an environmental reporter).[3] Reading the comments left by readers on that article led me to Brian Romans, a geologist who writes about island formation for his blog, *ClasticDetritus*.[4] It was his comment that led me to start thinking about plate tectonics.

The point of this is that following the comments in a blog is often a great way to find experts. (But be sure to check their bona fides.) There are lots of random people commenting authoritatively on blog streams as well. If you can't prove that they actually know something about a topic (usually because of a history of good writing on that topic), then quickly file under "don't know" and move on to people whose comments you can trust.

5. *Use context terms to limit the diversity of your search results.* When I was doing my searches, I found that adding the context term *geology* reduced the number of off-topic results and ended up focusing more on scholarly articles.

6. *Sometimes search requires ... well, searching.* Once upon a time, the verb "search" meant to spend the time looking for something. In this case, I had to hunt around a bit looking for a way to phrase my query to get something useful. Sometimes you get lucky and your first query gets you into a set of results that resolves the question. This kind of question is more complicated. For something like this topic, it might take a few probes to find a valuable set of results. Get in, check out the results, and if you don't see what you like, move on. But learn as you go; notice the terms used along with the publications and people involved. Usually that is enough to point you in the direction you need to go.

What started this whole thing was the wonderful map made by Raven Maps. I have no affiliation with Raven Maps other than being a fan. Here's the link to the map that sparked the question.[5] You have to see the map in person to get the full effect—but when you look at it, you'll immediately see that the right-hand side is VERY different than the left-hand side. The East Coast is full of detailed inlets, bays, and islands. The West Coast is relatively straighter, with fewer small details. Hence my original question.

Figure 8.7
Map of the United States showing the details of the East and West Coasts in high resolution.

Try This Yourself

This chapter began with an observation about something I noticed on a wall map, and ended with a long series of searches to learn about the differences between the East and West Coasts. Somehow, this ended up with a discussion of plate tectonics and sedimentation rates.

Noticing something and then figuring out WHY that something is that way often leads to fascinating research. Maps are particularly good at letting you see features of the landscape at different scales, and as we know, looking differently at the data frequently leads to insights.

For example, the other day I noticed that there's apparently a gentle arc of large lakes stretching across North America. You can see it in the following diagram (figure 8.8).

When you see such an apparently unusual alignment of lakes, you might think to yourself, How can I find out why this is so? (While keeping in mind that it *could* be simply random.)

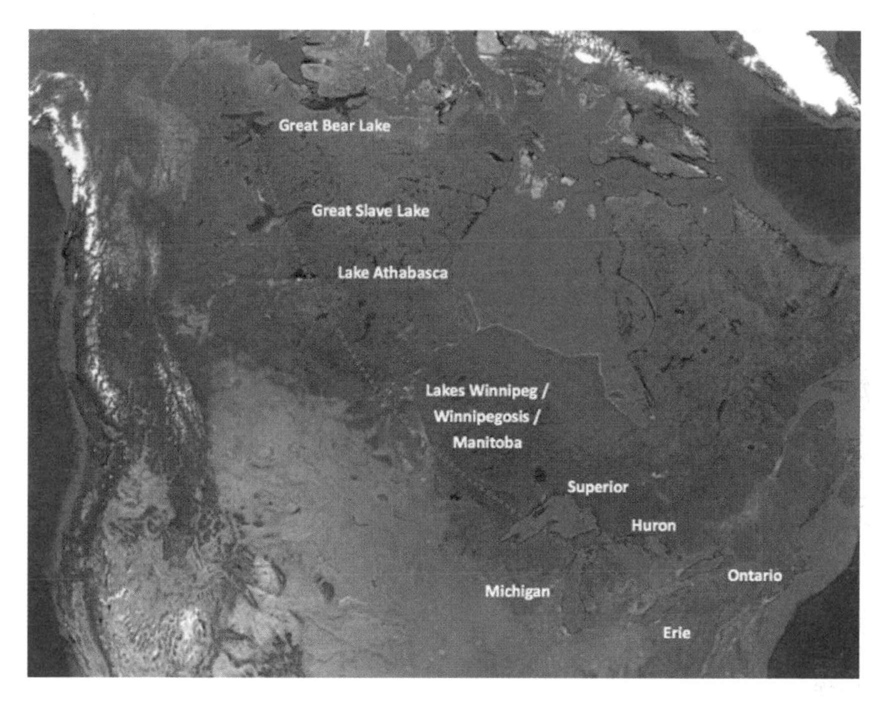

Figure 8.8
An unexpected line of lakes strung across North American raises a question: Why are they arranged like this?
Credit: Daniel M. Russell

How would you frame a research question for this? And now, knowing about how to ask good research questions, where would you start in answering such a profound one? (Here's a big hint: doing research starting with an observation might lead to deep questions about large-scale geographic features. Don't be afraid to follow your research clues to wherever they lead.)

9 Mysterious Mission Stars: How to Read Snippets in the Search Results and Pay Attention to Search Details

Every Spanish mission seems to have an unusual, beautiful, star-shaped window. Is this really a design feature of Spanish architecture from the colonial era? How can we find out?

I was recently in the seaside village of Carmel, California, visiting the old Spanish mission there. It really is a beautiful place; the courtyard is full of flowers, gentle sea breezes flow in from Monterey Bay, and thousands of hummingbirds zip from place to place. It's a traditional Spanish colonial mission building, with all the architectural stylings that you'd expect—domes, archways, and a distinctive four-lobed-star window. I've seen enough of these stars to wonder about them. Where did they come from?

To figure out the story lurking behind the star window at the main entrance to the old Spanish mission in Carmel, a picture is a helpful start.

I was wondering about that prominent star-shaped window over the arched doorway. It's pretty clearly a colonial era Spanish design element, but I'm not sure I've seen it at any other missions in California.

This prompted a few questions for me.

Research Question 1: *What is the story with that unusual star over the doorway at Mission Carmel?*

Research Question 2: *Does any other Californian mission have a star window like that?*

Research Question 3: *Where did that particular star shape come from? Can you find any other examples of stars in this shape in architecture elsewhere in the world?*

Figure 9.1
Above the door at the Carmel mission is a window of a distinctive design. This star pattern appears throughout colonial Spanish architecture. What's the story behind it?
Credit: Daniel M. Russell

I started with the first search I could think of:

[star window Carmel mission California]

The results I found were pretty interesting.

The first thing that you learn is the actual name of the mission: Mission San Carlos Borroméo de Carmelo.

Everyone just calls it "The Carmel Mission" or "Carmel," although just the single "Carmel" can also refer to the nearby small town that's famous for expensive real estate, expensive art shops, and the expensive people living there.

But note the two texts that I've circled on the SERP. The first tells me that there's a "Moorish" architectural element here (which I hadn't thought

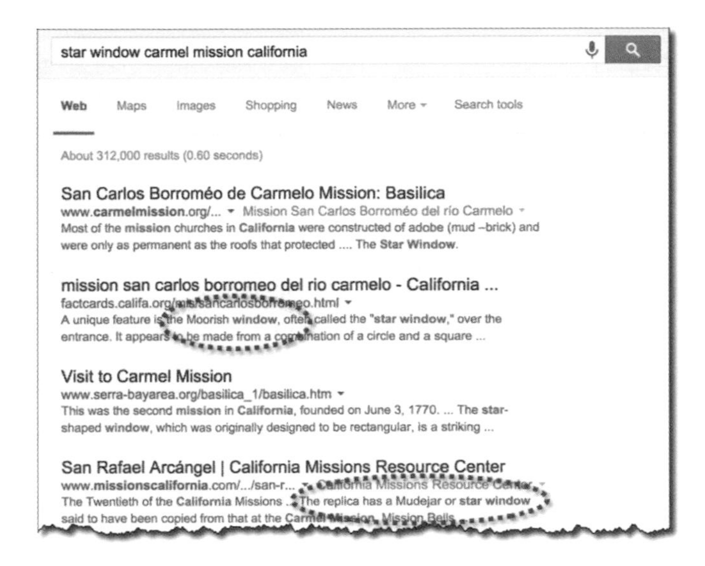

Figure 9.2
Just by looking at the search results page (SERP), you can learn a lot, but don't stop there!
Credit: Google and the Google logo are registered trademarks of Google Inc., used with permission

about before), and the second ellipse tell me that "the replica has a *Mudejar* or star window." That's a word I haven't seen before.

I read both those pages, and sure enough, I see that this entire mission was designed by Manuel Ruíz, a master mason from Mexico City who apparently also designed the Royal Chapel in Monterey (which is nearby, but is NOT the same as the village of Carmel). Ruíz is said to have incorporated Moorish elements into the plan, such as the domes, and the Moorish window, often called the "star window," over the entrance.

But as the second reference suggests, this kind of window is also called a Mudejar.

I do a quick search for:

[mudejar]

It teaches us that "Mudéjar also denotes a style of Iberian architecture and decoration, particularly of Aragon and Castile strongly influenced by Moorish taste and workmanship."[1] The Mudejares were Muslims who lived in Spain after the 1200s.

Figure 9.3
Mudejar stars appear all over Spain in artwork and even along the Autovía.
Credit: Daniel M. Russell

I then do an image search for:

["mudejar star"]

It shows lots of versions of the star in different locations in Spain (including two signs along the Autovía Mudéjar).

As that fourth result in the SERP above suggests, Mudejar stars also decorate other missions, such as the one in San Rafael, prompting another search for:

[San Rafael mudejar star]

In figure 9.4, Google Images shows us the star in the San Rafael Arcángel mission (which is about a hundred miles north of Carmel).

San Rafael Arcángel (1817) is one of the last and most northern of the Spanish missions to be established. It is unique among all the California missions in the fact that it was primarily established as a sanitarium to help heal the native population. (It was set up as a healthy place to go because Mission San Francisco de Assisi, in the current city of San Francisco, was so foggy and damp. This surprises absolutely nobody who has visited San Francisco in the past two hundred years.)

Oddly, when you do a search for San Rafael Arcángel in Google Maps, you see something different. Where's the white building with the Mudejar star? When you search for Mission San Rafael Arcángel, you land here—and while that mission building has a star window over the door, that's not *quite* what we're looking in this case!

Figure 9.4
A small Mudejar star over the entrance to one of the Mission San Rafael Arcángel buildings, but where is this building?
Credit: Daniel M. Russell

Where is it? Ah, with just a little **poking around in Google Streetview**[A], I found the "mission building" (it's the white building behind the palm tree in the above image). (See figures 9.5 and 9.6.)

THAT's where the San Rafael Mudejar star is!

But I know that the California missions all had a complicated history. So I search for:

[Mission San Rafael history]

That leads me to a number of sites that agree: the original San Rafael adobe church was built in 1818, but not long afterward began to disintegrate. San Rafael Arcángel was one of the first missions to be secularized, and in 1833, it was turned over to the Mexican government. By 1844, the mission had been completely abandoned. And somewhere between 1861 to 1870, the ruins of the mission were completely torn down and replaced by a new parish church.

Today, all traces of the early buildings have been lost and all that is left of the original mission is a single pear tree from the old mission's orchard. For this reason, San Rafael Arcángel is known as the "most obliterated of California's missions."[2]

Figure 9.5
Credit: Map data © 2015 Google

Figure 9.6
Credit: Map data © 2015 Google

Figure 9.7

Drawing of Mission San Rafael Arcángel around 1831 by Dr. Eduardo Vischer, bit. LY/TJOS-9-1.

Credit: Library of Congress, Prints and Photographs Division, HABS CA-1131

Luckily for us, on December 18, 1949, a replica of the original mission church with a simple doorway under a star window was dedicated. (I tried to find where this particular design came from, but I haven't yet succeeded. Is the "replica" really a replica? Or is it just kind of like the original?)

From MissionTour.org we learn that in 1878, General Mariano Vallejo (the governor of California at the time) approved a drawing of the mission done by Edward Vischer (a German-born artist who made a living drawing missions and the Californian scenery). MissionTour.org says that it found this drawing in the collection of the Historic American Buildings Survey run by the Library of Congress. Here is a poorly digitized 1940 photo of that drawing (figure 9.7).

That's an interesting lead. So I visited the Historic American Buildings Survey site and did the search for [Mission San Rafael]. The Vischer drawing is from this Library of Congress collection. Notice the shape of the window at the end of the building over the door; it's fairly square.

Yet THIS painting by Renaud (a contemporary of Vischer) dates from before 1835, but how far before is unknown. Here it's Mudejar star-shaped!

Figure 9.8
San Rafael mission, painted by Renaud, sometime before 1835, bit.LY/TJOS-9-2.
Credit: Library of Congress, Prints and Photographs Division, HABS CA-1131

And to complicate things, here's another by Oriana Day (figure 9.9), who painted all the missions between 1861 and 1865.

If I had to guess, I'd say that Day was taking a few liberties. Note that the other two paintings clearly show a walkway with many posts supporting the roof, while the Day portrait does not.

In any case, there's a Mudejar star in the "replica" building in San Rafael now.

To find other examples of the Mudejar star is fairly straightforward. As we saw in the earlier illustrations from Spain, they're easy to find with Mudejar as an additional term in the query. But naturally, I wanted more. (Or perhaps, I wanted Moor proof.)

So I queried:

[Moorish architecture Mudejar star]

That search gave me a LOT of examples of stellate images from the Alhambra, that quintessentially Moorish palace and fortress complex located in Granada, Andalusia, Spain.

But as I entered the query

[Alhambra mudejar …],

Figure 9.9
Another version of the mission, again with a square window over the door, painted by Oriana Day between 1861 and 1865, bit.LY/TJOS-9-3.
Credit: Library of Congress, Prints and Photographs Division, HABS CA-1131

I started to see some interesting auto-completion suggestions. The second suggestion told me something useful: *arte* is a Spanish word. (See figure 9.10.) Maybe I should try searching in Spanish. This is, after all, an inherently Spanish term, and there's probably a lot that I'm missing by searching in English.

The simplest way to **search in Spanish**[B] is just to translate the query into Spanish, and then enter it into the search box and see what comes out. If I translate "star" into Spanish, it's *estrella.* That query comes up with the following.

In this case, a query with estrella gives us even more fantastic examples of Mudejar stars/estrellas (figure 9.11).

Although I have to admit that my favorite example, which I found just by browsing through the images, was this beautiful fountain in the Alhambra (figure 9.12).

This fountain pretty neatly connects the "star window" in the San Carlos Borroméo de Carmelo Mission in Northern California with the Moorish designs of the Alhambra from the heart of Spain.

Figure 9.10
Suggestions that appear when the query is typed into the search box.
Credit: Google and the Google logo are registered trademarks of Google Inc., used with permission

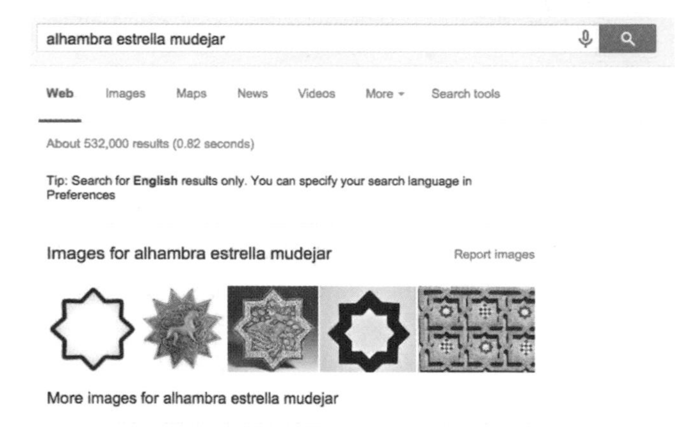

Figure 9.11
Sometimes searching in another language (in this case, the obviously appropriate language) can give insight into your search process.
Credit: Google and the Google logo are registered trademarks of Google Inc., used with permission

Figure 9.12
This version of the Mudejar star is found in the classical palace known as the Alhambra in Granada, Spain.
Credit: Juan R. Regaldie, jrphoto.regaldie.com

Research Lessons

Starting from the top, we learned a few things.

1. *You can learn a lot by reading snippets*. But as I've written before, be careful not to overread the snippets. You can get leads, but follow them up so you understand what's really going on. In this case, we got our first idea that this could be Moorish in origin and called a "Mudejar star."

2. *Reading results gives you clues*. It was by reading about Mudejar stars that we first found the hint that San Rafael Arcángel might have a star window as well. Of course, that led us to the historical quandary about whether or not it originally had such a window. Nevertheless, we found it.

3. *Look around in Google Street View when you can't find the thing that everyone says is there*. In this case, the Mudejar star window wasn't obvious—until you "walked" pegman down the street and looked back.[3] THEN it was clear where the star was.

4. *When searching for Spanish things, try searching in Spanish!* This is obvious in retrospect, I know, but I finally figured it out.

5. *Notice what auto-complete is trying to tell you.* In this case, one of the auto-completions was in Spanish. Pay attention. Read the signs along the way.

6. *Pay attention to the world around you and let your friends help out.* I have to point you to the excellent podcast about "The Fancy Shape" on the 99percentinvisible.org website.[4] This particular podcast is dedicated to the quatrefoil, a four-leaf clover design, which when you add little barbed-shaped triangles on each lobe becomes the Mudejar star. Oddly enough, I found this podcast purely by serendipity. I'd mentioned my interest in the Mudejar star to a couple of friends, and sure enough, just a few days later, one of them emailed me this link to the podcast. Serendipity can be amplified by social networking and letting people know of your interest. This casts a broad net on your behalf, and suddenly, you've got dozens of eyes and ears scanning the world around you for your research questions. It pays to have friends.

How to Do It

A. Poking around. An incredibly useful thing to know for using Google Maps is how to just "poke around" using Google Street View. Here's the way that Google Maps looks when you've just done a search for [San Rafael Mission] (depending on where you are, you might have to add California in the query).

Now to look around in the area, you can use Google Street View (a collection of images taken on the streets). To see what's covered in the area, just click on the little yellow icon of a man in the lower-left corner. That's the *pegman*, which we'll use to explore the area. Once pegman is clicked, blue lines will pop up on the map indicating all the streets that have Google Street View images. Notice that there are also some small blue circles. Those are photospheres, which are 360-degree immersive photos you can use to explore an area that's not on the street (figure 9.13).

To explore an area, click and hold the pegman, dragging him over to some part of the blue lines (or a photosphere circle). When you release the pegman, the image will change to show you that part of the Google Street View.

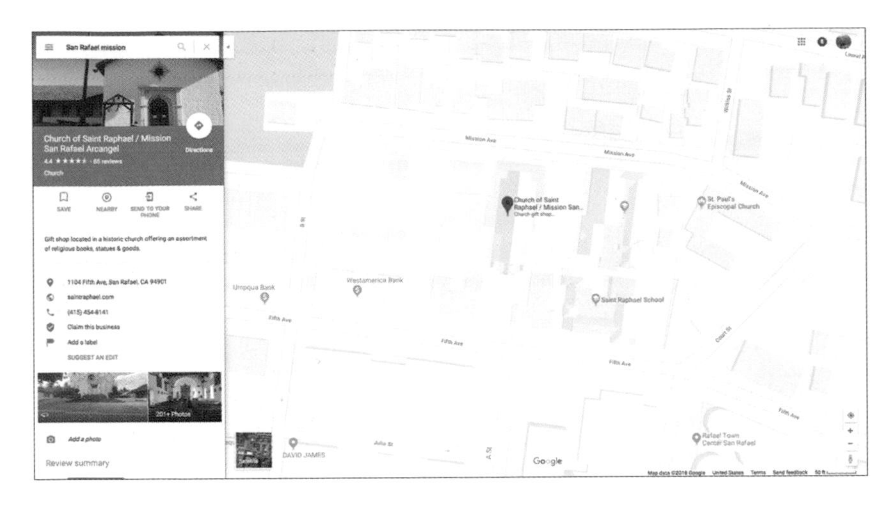

Figure 9.13
Searching for [San Rafael mission] in Google Maps takes you to a browsable map of the area around the mission.
Credit: Map data © 2018 Google

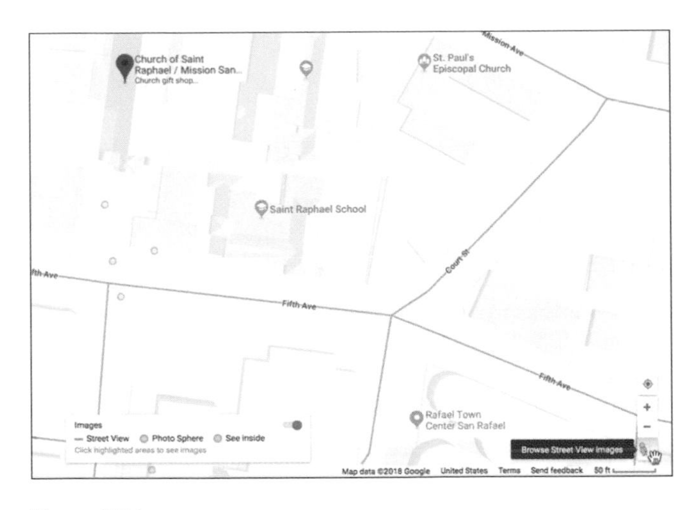

Figure 9.14
Once you click on the pegman, blue line and blue circles appear showing where you can explore the Google Street View imagery or get a 360-degree immersive view in a photosphere.
Credit: Map data © 2018 Google

Figure 9.15
The Google Street View scene in front of the San Rafael mission. From this location, you can see two windows with star shapes.
Credit: Map data © 2018 Google

Once you've found a location that you want to explore further, you can then press and hold on the Google Street View image, and then drag it back and forth to see all around. Once you've found an item of interest (say, that star-shaped window in the building on the right), just click on the + symbol in the lower right to zoom in (figure 9.16).

Finally, to move your location on the street, just pivot your view to look at the street, and then hover your mouse over the street to see an arrow appear on the street. Clicking on the arrow will move your position to that point on the street. That's how you poke around in an area to get different views of the thing you're searching for—in this case, a better view of the star over the window on the building to the right (figure 9.17).

B. Search in Spanish. With online translation services, searching in other languages has become fast and simple. When I want to search in another language, I just search for [Google Translate], which brings up a translation box at the top of the SERP:

Then you can copy/paste the word(s) from the right side of the tool. (And notice that you can click on the little speaker icon to actually hear the translated text spoken aloud, which might be handy.) At this time, Google

Figure 9.16
You can zoom in or out of a Google Street View image by clicking on the + symbol in the lower right.
Credit: Map data © 2017 Google

Figure 9.17
To actually change your Google Street View location, you can hover over the road until the arrow appears. Clicking on the arrow changes your location to that place, allowing you to move around to change your viewpoint.
Credit: Map data © 2017 Google

Figure 9.18
By searching for [Google translate], you can get to a fast translation service.
Credit: Google and the Google logo are registered trademarks of Google Inc., used with permission

Translate can handle over a hundred languages, from Albanian to Zulu, so should you need to search for a special term in Zulu, it's really fairly easy.

Or if you just want to translate a single word or short phrase, you can do a query like this:

[star in Spanish]

But notice that you can do more than a single word at a time; you can paste large blocks of text, which then will be translated.

Try This Yourself

Just plain old poking around can often lead to remarkable discoveries—especially since you can now poke around using Google Street View. When you combine the ability to wander around another country and Google Translate to change the text you see into your own language, you can visit another country and do lots of research from your computer.

For example, touring Rome, Italy, via an online connection can lead you down some fascinating alleyways.

If you look for the Google Street View images of the Colosseum (aka *Colosseo* in Italian), you can wander through the interior of this giant battleground that's nearly two thousand years old.

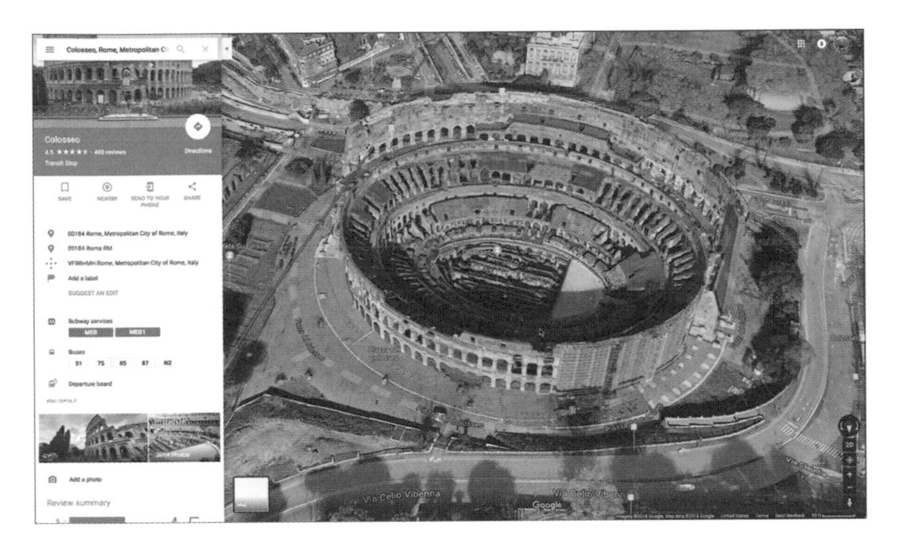

Figure 9.19
Using Google Maps lets you wander around the Colosseum in Rome. Google Street View lets you walk through the interior of the building and explore the side streets. Credit: Google and the Google logo are registered trademarks of Google Inc., used with permission

Once you find the Colosseum, see if you can find the Domus Aurea nearby. With Google Translate, you'll be able to understand what the side street names really mean. As is the case with many richly historic cities, street names often capture a good deal of history.

What kinds of things can you discover by just wandering through the side streets around the Colosseum? (Careful: once you start virtually strolling through Rome, you might have to visit in person.)

10 When Was Oil First Discovered in California? How to Discover and Work Through Multiple Competing Claims in Online Resources

How can you get to the bottom of a simple question that has many different answers? Watch out for different versions of the same story!

I grew up in Los Angeles, living somewhere between the beach and Hollywood, in that part of the city that was covered in oil wells, pumpers, and derricks that were drilling for even more oil.

You might not think of Southern California as a big oil production region, but there are oil wells *everywhere*—from the northern coasts of Santa Barbara down to the long white beaches of San Clemente. The petroleum fields even extend way up into the long, rich agricultural areas of the Central Valley. Not only does the Central Valley produce most of the fruits and nuts for the United States, but it has gigantic oil deposits.

Growing up, for many years I didn't think much about how, why, or when oil was first produced in the state. I just assumed every place had its own oil supply. Didn't every place have oil derricks and pumpers going up and down all day, every day? It wasn't until I moved out of California that I realized oil isn't everywhere, and it wasn't always there, like agriculture.

Realizing this, I started to wonder, when and where was oil first discovered in California?

I began doing a bit of searching, and was surprised at how hard it was to get a clear and verifiable answer to a simple question. After all, shouldn't it be fairly clear WHICH oil well was drilled first, and WHERE the oil was processed? Wouldn't a quick Google search just answer the question?

Figure 10.1
Grasshopper-style oil wells in the Southern California sunset.
Credit: Daniel M. Russell

This led to my first research question.

Research Question 1: *When and where was oil first discovered in California?*

When I initially started working on this, I did the obvious search query:

[California oil discovery]

I found lots of resources, each willing to tell me where oil was first found, and a few willing to tell me where it was first refined (or "distilled" or "rectified," depending on what kind of person was writing the text and when the text was written).

One of the first places I found was an article titled "What's with Oil? The Probable Status of the Resource." It was published by the Redlands Fortnightly, a "paper reading" club that's been active in topics social, agricultural, and historical since 1895. I thought an organization with that amount of history would have a useful perspective.

In that document, the author writes,

> In 1865, only 6 years after "Colonel" Edwin Drake's monumental discovery in Pennsylvania, California's first productive well was drilled by the Union Mattole Company in California's Central Valley. This area, east of San Francisco, became

> the scene of much of the drilling activity through the rest of the 1800's. While none of these wells were considered major strikes, they did provide enough oil for the nearby market of San Francisco, by far the largest population center in California in the late 1800's.[1]

The author is Ty Schuiling, who has degrees in geology, and has worked in the oil industry and at USGS for twenty years.

This would seem conclusive. The only problem is that the Union Mattole Company was active in the Mattole River area. Why is that a problem? When I read that, I realized that I had no idea where the "Mattole River" is—so I looked it up using Google Maps. To my surprise, it's in Humboldt County, which is fine, but it's nowhere NEAR the Central Valley or Southern California. This isn't a minor error; it's off by a few hundred miles. (For the curious, the Union Mattole Company operated out of the town of Petrolia, previously known as Petrolea—a fact that complicates the search.) This kind of error calls the whole reference into question. Yes, the Central Valley IS a major oil area now, but it wasn't in the 1860s. How could the author make such an odd mistake?

Take note of that phrase, "*California's first productive well was drilled by the Union Mattole Company in California's Central Valley.*" If you do an **exact phrase search**[A] on this, you'll find it's used in twelve different documents. Hmmm. Really? How is that possible?

Another claim that's made in other documents is that oil was first discovered in Southern California near Santa Clarita at the Pico No. 4 site. The problem here is that most of THESE articles use the phrase, "*Many people may be surprised to learn that one of Southern California's chief exports over the last 100 years, besides motion pictures, has been oil.*" If you do a search for the quoted phrase ["one of Southern California's chief exports over the last 100 years"], you'll find it's used in thirty-seven different documents. What? The exact same phrase?

Something fishy is going on here. Why so much obvious copying? I spent several hours looking through all kinds of articles, websites, original news archives, and books to find out. One thing became strikingly clear: there is a HUGE amount of text reuse and repetition on this topic. This makes it difficult to figure out what actually counts.

So after lots of searching through the different documents I found online, what did I discover, and how?

The first thing I noticed is that people knew about the presence of oil, tar, and asphaltum (a kind of semisolid oil) in California for a long time. For instance, Carpinteria was a village to the Chumash people, who used the naturally occurring asphalt to seal their canoes, known as *tomols*. Petroleum seeps can be seen (and found on your feet when you walk on the beach) near the beach bluffs at Tar Pits Park by the campground of Carpinteria State Beach. This was noted by Fray Juan Crespi in 1769 when his exploration team visited the area.[2]

This tells me that we have to clarify our research question just a bit in order to get a better answer—one that we can definitively answer. Here's the new version of the research question that makes it clearer.

Research Question 1 (revised): *When was oil first produced as a commercial product in California?*

There seems to be four major claims on all the web pages that I found on this topic. From my reading notes they are as follows.

1. *In 1876, oil driller Charles Alexander Mentry* (aka Charles Alexander Mentrier) struck oil at Pico No. 4 in Pico Canyon in 1875, but a gusher drilled in 1876 was the first truly productive well, which continued to produce until 1990.

2. *In 1865, the Union Mattole Company* struck oil in Humboldt County and shipped several barrels to San Francisco for refining.

3. *Edward Doheny in 1892* used a sharpened eucalyptus tree trunk to drill down 150 feet near present-day Dodger Stadium in downtown Los Angeles.

4. *In 1855, Andreas Pico* distilled small amounts of oil from Pico Canyon, but had a limited market.

After finding each of these claims, I did a search to determine the origins (and repetitions) of each. As mentioned, the repetitions are numerous; getting down to an original claim is surprising difficult.

The evidence for each is as follows.

1. *Mentry*: Chevron's corporate history notes that "in September 1876, driller Alex Mentry succeeded in striking oil in Pico No. 4, despite rattlesnakes, wasps, mud and underbrush. The first successful oil well in California, Pico No. 4 launched California as an oil-producing state."[3] The oil from Pico No. 4 was successfully refined in that year, according to Chevron

(which ought to know, as Pico No. 4 was its well, inherited from Standard Oil of California, its predecessor company). The well continued to be productive for another 114 years until it was capped in 1990.

As a major oil company, however, Chevron has a vested interest in having "the first" or "the biggest," so while this claim is true enough, it depends on what "first successful" means.

Curiously, Pico No. 4 is famous enough to have its own Wikipedia entry.

From the Wikipedia article on "Pico No. 4," we learn that "well No. 4, the Pico Canyon Oilfield, located about seven miles (11 km) west of Newhall, California in the Santa Susana Mountains, was the first commercially successful oil well in the Western United States, and is considered the birthplace of California's oil industry. Drilled in 1876, it turned nearby Newhall into a boomtown and also spawned a smaller boomtown called Mentryville adjacent to the drilling site."[4]

By following up various leads, it's easy to find other references that support this assertion. For instance, in 2003, Nicholas Grudin, a staff writer at the *Los Angeles Daily News*, in his article "Ghosts of an Era: Mentryville Is a Monument to Both the Start and Decline of the Area's Oil Drilling Industry," wrote that "Scofield formed California Star Oil Works, and with skilled oil man Alex Mentry, tapped the first commercial oil well in California—Pico No. 4."[5]

There's also Jonathan Gaw's 1993 article in the *Los Angeles Times*, "Oil in a Day's Work: The Boom May Be Over, but a Few Wells Pump On," in which he writes, "Oil men had been groping around the canyons of the area since 1876, when the first commercially successful oil well west of Pennsylvania was built several miles south of Lechler's ranch in Pico Canyon."[6] (Of course, that would be Pico No. 4.)

2. *Union Mattole Company*: Wikipedia, in an article on the history of oil in California, goes with the Union Mattole Company find on the Mattole River in 1865.[7] It then amplifies a bit on "first productive" well with a slightly contradictory comment about the Brea-Olinda oil field in Southern California that came online in 1880. (The reference for this claim is Walter August Ver Wiebe's piece in the 1950 book *North American and Middle Eastern Oil Fields*.)[8] On the other hand, the California State Parks website says that "California's first drilled oil wells that produced crude to be refined and sold commercially were located on the North Fork of the Mattole River

approximately three miles east of here [the town of Petrolia]. The old Union Mattole Oil Company made its first shipment of oil from here, to a San Francisco refinery, in June 1865. Many old well heads remain today."[9]

I also found in the *Geology of Southern California* that "the first drilling for oil in California probably dates from 1861, when a prospect well was drilled on the Davis Ranch in Humboldt County in the northern part of the State. The first recorded production of crude oil in California was from a well drilled by the Union Mattole Oil Company in 1865, near the Mattole River in Humboldt County. The depth of this pioneer oil well is reported to have been 260 feet." Take note, however, that in the same document, a "prospect well was drilled in 1861 on the Davis Ranch in Humboldt County."[10] Apparently, the prospect of oil on the Davis Ranch also didn't work out.

If you continue searching in Google Books for Mattole and oil, you'll find multiple sources that argue the Mattole River area was the first commercial oil site in the state. For example, a much-earlier publication, the *California Journal of Mines and Geology* (from 1883!), says pretty much the same thing.[11]

Another book from that same search process, *Early California Oil: A Photographic History, 1865–1940*, maintains that "the birth of the California petroleum industry proper may be dated to March 25, 1865, with the first commercial sale of oil refined in the state. [T]he Union Mattole Oil company was incorporated [on] June 7, 1865, [and] the first shipment was sent to San Francisco where it was distilled by the Stanford brothers."[12]

It's worth knowing that Walter Stalder records that the Stanford Brothers refined and sold the first shipment of oil from the Mattole well, making it the first oil produced and refined from a California well.[13] Reportedly, according to Stalder's article, "Contribution to California Oil and Gas History," the refined "burning oil" sold for $1.40 per gallon.[14]

I found all these sources just by searching inside Google Books for variations on **Union Mattole**, **oil**, and **commercial**. (You could imagine using other terms, but these were productive in terms of finding resources.)

This last book resource is a really useful discovery. Not only does the 1865 Mattole claim have multiple sources of support, but finding a price for the refined product lends a good deal of credibility.

But I can't resist adding in this tidbit from another article that I found, written by a local historian in the Mattole River valley:

1861: Discovery of oil in the Valley first publicized.

1864: All but a dozen or two of the least troublesome Natives killed or captured. Indian troubles considered over. In 1868 measles kills most survivors.

1865: First oil shipped out by Union Mattole Co. Principal town established and named "Petrolia." Oil boom short-lived, though experimental drilling and subsequent oil excitement recur periodically.[15]

Just to muddy things up a bit, an article in the *Humboldt Times* from 1907 claims that "several large companies began extensive oil operations in 1898." The author goes on to say that oil will cause the Mattole River valley to flourish, opening up ever-larger markets for the export of tan bark (which she clearly thought would be the salvation of the region, yet with the benefit of twenty-twenty hindsight, we can see how wrong she was).[16]

3. *Doheny*: The Doheny claim for commercial oil production in 1892 is made in the book *Petroleum in California: A Concise and Reliable History of the Oil Industry*.[17] But I have to admit this seems a bit unlikely to me. It has all the hallmarks of a tall tale. A well was drilled 150 feet deep with a 60-foot (or 20 meters) eucalyptus tree trunk? Despite the title of the book, I find it difficult to believe this story, although the text ("struck oil by using a 60-foot eucalyptus") does get repeated often in different versions of the Doheny story, in exactly the same copy-and-paste language.[18]

4. *Pico*: Interestingly, this claim of early oil refining by Andreas Pico is ALSO made in the book *Petroleum in California* and a few other places, but all the articles agree that this was small-time distillation, just for the kerosene lights of the mission in San Fernando to use as illumination. While successful, you wouldn't really call it a commercial success.

Of course, looking at all these claims more or less inexorably took me to the Wikipedia entry on the "California Oil and Gas Industry," which claims that oil has been *known* since the dawn of time from the large number of oil and tar seeps that seem to sprout from the soil every few miles throughout the state.[19] There are oil seeps in Northern California, the Central Valley, and of course Southern California. But the *Petroleum in California* book tells us that a company began working the La Brea tar pits (in Los Angeles) in 1856 by building tunnels that would let the oil run out into holding ponds, although there's no reference cited. This seems fishy to me; tar pits are on pretty flat land, and the closest hills are the Hollywood Hills, which are three miles from the pits. How to check up on this claim? I did a quick check with the query:

[1856 oil gas history tunnels]

I was able to find a document *Oil and Gas Production: History in California* (published by the California State Department of Conservation).[20]

This was a huge surprise; yes, people DID dig tunnels at the La Brea tar pits near the exact center of Los Angeles, but gave up when drilling holes in the ground proved much simpler and more productive than digging tunnels and letting the oil flow out slowly.

Some thirty miles (or forty-eight kilometers) as the oil-black crow flies from the tar pits, near Santa Paula, California, Josiah Stanford dug about thirty tunnels into the oil-bearing mountainsides, slanting them upward so oil flowed down to the entrance. Some tunnels reportedly produced up to twenty barrels of oil per day. The oil flowing steadily from the tunnels made Stanford one of the top oil producers of the late 1860s. The tunnels produced more oil in California than any other method until the development of oil wells drilled with more modern methods. These tunnels continued to slowly deliver oil until the mid-1990s, when they were finally abandoned and plugged up. (Strangely enough, another thirty miles [or forty-eight kilometers] due west of Santa Paula are the still-productive oil islands in the Santa Barbara Channel. One of these—Platform A—spilled around a hundred thousand barrels of oil in early 1969, making more than a few seabirds and crows truly oil black.)

My conclusion from looking at lots of evidence for the competing claims is this: the Union Mattole Company had the first commercial sales of refined oil from a California oil well in Humboldt County on the Mattole River in 1865. Alas, the well wasn't productive for very long (a year or two at most), and so the award for earliest, best, long-term continued success goes to Pico No. 4, which began operations eleven years later in 1876, with continuous oil production for over a century.

As we've seen, it's not simple to figure out when and where oil was first discovered in California. In fact, it's complicated for two key reasons.

First, there's a lot of not-so-great scholarship out there, even in books and what seems like primary sources. In particular, there's a lot of repetition of errors that were introduced early on and somehow never corrected.

Figure 10.2

California's first commercial oil well, Pico No. 4, produced thirty barrels per day when Alex Mentry completed it on September 26, 1876, for the Star Oil Works Company. This prompted the owners to set up the Pacific Coast Oil Company, a forerunner of Standard Oil Company of California.

Credit: Star Oil Works, San Fernando District, S. P. R. R., photCL 74 (502), Carleton E. Watkins photograph collection, Huntington Library, San Marino, California

(That's what I mean by poor scholarship. Just repeating something that you found on the internet doesn't make it right; it merely makes it repetitious. And if you didn't double check it in the first place, you're earning yourself a particularly nasty place in hell.)

Second, there are multiple interests at work, each arguing that THEIR particular oil strike was first. You can imagine why they'd say this; being first to find oil gives you a special cachet, a kind of romantic wildcatter kind of image.

Research Lessons

As you do your searching to find answers to your research questions, you're usually seeking something that's rare. But every so often, you'll find too much information from different sources that's just a bit inconsistent. In this case, you need to sift through what you find. Here's how to do that sifting in order to find what's true and accurate.

1. *You need to know WHAT it is you're searching for and define the question carefully*. What does it mean to be the "first to discover oil in California?" If you consider that the Native Americans had been using tar from the La Brea pits, then the *discovery* was when they first showed up on the scene, back in the days of saber-toothed cats, mastodons, and dire wolves (around 10,000 BCE). That led me to shift the research question a bit to make it something that I could definitively answer.

2. *Be aware of duplications in the content you find*. There are many reasons you'll find duplicated content, but when the duplications are unattributed, you have to be a little wary about the content. It might be correct, and the duplication is just careless editing, but it also suggests that the authors are not paying particular attention to their writing. As a side effect, texts that have lots of duplicated content don't really count as another citation. It's just copied.

3. *Organize what you find*. This turned out to be a complicated research question because there's SO much content, with much of it telling variant stories. In the end, I started taking notes on each of the competing claims, with one set of notes for the Union Mattole claim, another for the Doheny claim, and so forth. All I did was to write down the key points (year of discovery, year of first commercialization, and so forth) and then the URLs

to the place where I found each assertion. In the end, I had a big table of competing claims, and the evidence to back them up. Armed with that set of notes, I could figure out the answer to the research question!

How to Do It

A. Exact phrase search. To search for a complete phrase, just put the phrase in double quotes. In the example above, I did this search:

> ["California's first productive well was drilled by the Union Mattole Company in California's Central Valley"]

Obviously, you want to get the whole phrase, and when you do, you'll see that this exact phrase is used verbatim on a number of different web pages from different sites, each claiming that this is the correct history of oil discovery in the Golden State. Often I'll use just a few distinctive words rather than a long sentence. In this case, the phrase "California's first productive well was drilled" would have been enough. (I used the entire sentence here because I was looking for long sections of duplicated text.)

Try This Yourself

This chapter is about ways to notice when a "second source" isn't really a second source—when it's merely a copy of another, earlier document. There are lots of reasons why writers do this; sometimes they're just copying the text as a reference, but sometimes they're just lazy and copy big pieces of text without thinking.

When you're doing your own online research, you have to take note when your sources are copying other sources. When you find the wholesale lifting of texts, you should be worried, since it suggests that the research underlying the document hasn't been checked carefully. Copying like that is the mark of a lazy researcher; don't trust that article (but perhaps look around for the original).

As I do my research, I'll sometimes notice when particular phrases (especially clever and curious turns of language) keep reappearing as I read. *Those* are the sentences (or sentence fragments) you want to double check.

One hint is to look for those repeated phrases. As you check for duplication, you don't want to check for duplicate titles. But a well-chosen phrase (that is, one that's central to the argument, and long enough to be unlikely to happen by accident) can be a useful way to see how far that article has spread. People often choose to copy rather than rewrite the central idea of an article.

For instance, in an article about the worldwide effects of the Chicxulub asteroid impact and how it caused a global mass extinction, a key sentence fragment is this: "lead us to conclude that the Chicxulub impact triggered the mass extinction."[21]

I chose the particular part of a sentence that summarized the entire article, but that wasn't so long it might be changed in the process. Do a search for this sentence fragment:

["lead us to conclude that the Chicxulub impact triggered the mass extinction"]

You'll find many other articles that include this sentence fragment. Check out the other articles to see if you can find how many of them reuse many of the ideas and some of the language of the original article. Can you find the original article to use that phrase? Search around and you'll be able to spot other key phrases that are duplicated across articles on the Chicxulub meteor strike. (And you'll be able to use this same duplicate-detection trick to check on topics that you're researching. Remember: just because you get many hits on a topic, if they're all textual duplicates, the additional references don't actually add any value!)

11 Can You Die from Apoplexy or Rose Catarrh? How to Find (and Use) Old, Sometimes-Archaic or Obsolete Terminology

How does language change over time? We've changed the names of a few important things over the past hundred years. Here's how to do your online research and not get confused when things change names.

One of the greatest problems with the past is that the language they spoke then is different than what we speak (or write, blog, or inscribe on our cars) now. You'd think that English is English, but the truth is that there's a fairly high rate of turnover in the language. New words enter, and old ones leave, but there's also a shift in the meaning of some words with time. (Depending on how old you are, "bad" means, well, *bad*, but in the mid-1940s, the 1960s, and again in the 1980s, "bad" had a slang meaning of "good!" Go figure. I have no idea why "bad" meant "bad" in the 1950s.)

As a consequence, when you're trying to search for historical content, you sometimes (often?) have to shift your language to accommodate the way that authors in the past would have written.

For instance, when I was young, I grew up learning about a dinosaur that was then called the brontosaurus. The way these things go, the name was more or less rescinded in favor of *apatosaurus*, and then brought back a few years later when a newer, finer distinction was made between the brontosaurus and apatosaurus.[1] If you're curious, you can go read that article to get the whole story about the name changes (and why they keep changing the name back and forth). It's an interesting story about science, with the scientists wanting to be clear about what they're studying.

But the reason I bring it up here is that *terms can change significantly over time.* One dinosaur name goes away and then comes back. If you think

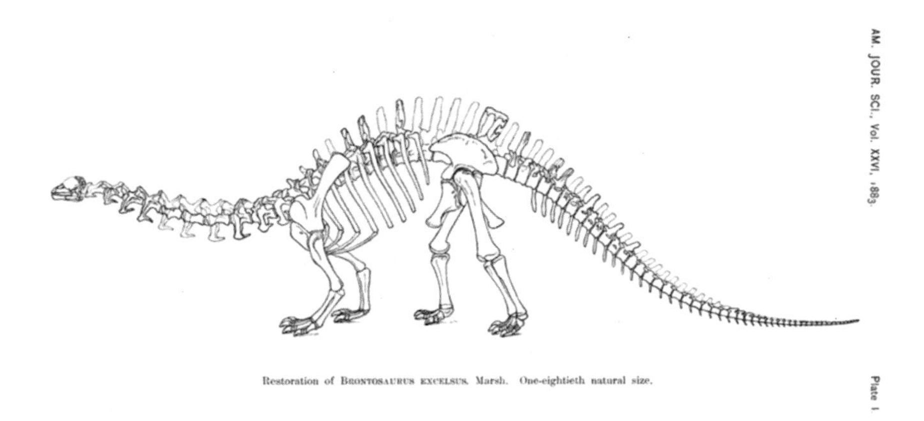

Figure 11.1
The named, unnamed, and restored (with a different skull) brontosaurus.
Credit: Reprinted courtesy of the *American Journal of Science* 5, no. 26 (1883)

about it, a lot of our language has shifted. “Boers” became “Afrikaners,” people who were “insane” became “mentally ill,” and the phrase “outdoor relief” became commonly known as “public welfare,” and so on. Of course, this is particularly useful to know if you're trying to do online research, especially about historical ideas.

This terminological shift (a great phrase to use at your next party!) showed up a bit in my own research recently. Here's an example.

I was doing a bit of reading about the US Civil War (1861–1865) and had read in one source that many of the soldiers died from some kind of disease that had extensive diarrhea. Given the sanitation conditions at the time, that's not surprising. What WAS startling was the claim that more soldiers died of disease than from actual combat. I know that around 2 percent of the entire US population perished during the war (that is, around 625,000 souls). How much of that was due to disease rather than direct fighting? Yet when I search in writings from that time, I find lots of diarrhea, but I seem to be missing many of the references that would help me figure this out.

Research Question 1: *Is there some other term (or terms) I SHOULD be using to search in archival accounts from that period for this disease?*

There are several ways to find terms for medical conditions used during the US Civil War. Here are a couple of methods.

Search explicitly for synonyms of terms that you know about. That is, start with a specific disease you can name, and look for uses during that period:

[diarrhea during the us civil war]

This leads us to a few documents that give us several synonyms for diarrhea: dysentery, quickstep, flux, and "alvine flux" by the doctors. Unfortunately, doctors knew neither how soldiers contracted the condition nor how the diseases should be treated. Likewise, do a search for:

[diarrhea synonyms]

It will give us a bunch more synonyms (Delhi belly, flux, Montezuma's revenge, runs, trots, and turista), but for each of these you have to check to be sure they're Civil War period relevant. For example, looking at the Google Ngram chart for "Delhi belly" shows that it didn't come into use until the late 1940s.[2]

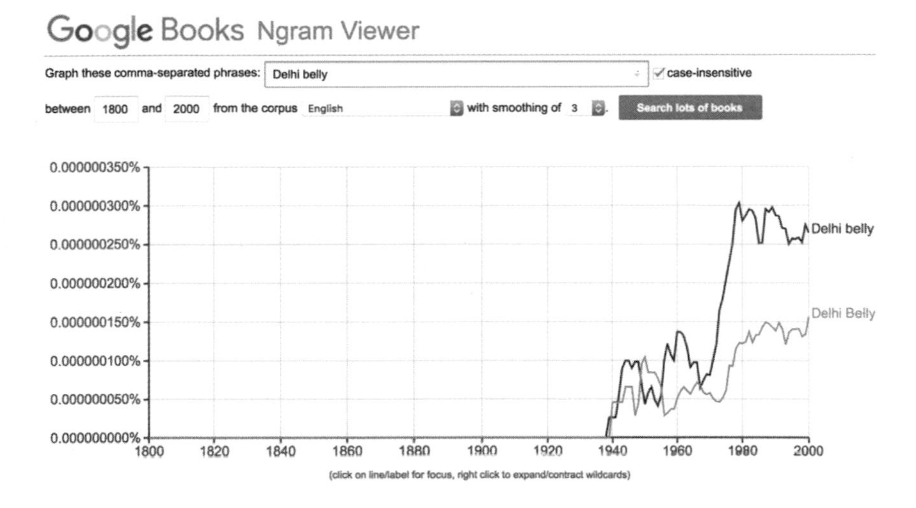

Figure 11.2
The term "Delhi belly" didn't come into common use until the 1940s. This chart is made using Google Ngrams, which shows how often a word (or phrase) occurs in the Google Books collection over the past two hundred years.
Credit: Google and the Google logo are registered trademarks of Google Inc., used with permission

Unfortunately, checking Ngrams for "flux" doesn't work because it's a common industrial term as well. We want only the interpretation of "flux" as it applies to our research. So we have to confirm that "flux" is a Civil War era term by doing a simple query:

[flux Civil War]

This leads us straight to a glossary of Civil War medical terms and confirmations that "flux" was a common term for diarrhea.[3]

For these kinds of historical topics, it's worth checking Google Books with a search like:

[Civil War language disease]

That leads us quickly to *The Language of the Civil War*, confirming "flux" and "bloody flux."[4]

The disease problem was massive. Soldiers had to contend not just with the fighting but also with typhoid, pneumonia, measles, tuberculosis, and malaria. An indication of just how bad this was during the war is shown in this chart from the Civil War Trust (figure 11.3).[5]

As you can see, MOST of the deaths during the American Civil War were due to disease; there were many terms, but any diarrheal problem was often called flux, quickstep, or trots. Those (along with regular old diarrhea, in all its misspellings, which were common at the time) are the search terms to use

Of course, not all disease is fatal, but when you're reading texts from the late nineteenth century, you'll frequently come across language that's archaic or just plain puzzling.

For instance, while reading about the death and destruction of flux/influenza, I happened across an unfamiliar term: *rose catarrh*. When I did a search for [rose catarrh], the definitions were a fairly unhelpful "rose cold." Really? That's it? So I kept at it, looking further and further down the list of results until I finally found a slightly more useful definition: "*a variety of hay fever, sometimes attributed to the inhalation of the effluvia of roses*." (This was from the 1913 edition of *Webster's* dictionary; you can tell because the word "effluvia" has such a nineteenth-century feel to it. Of course, my first reaction was to do a quick definition search on Google: [define effluvia], which tells me that it's "an unpleasant or harmful odor, secretion, or discharge, such as 'the unwholesome *effluvia* of decaying vegetable matter.'"

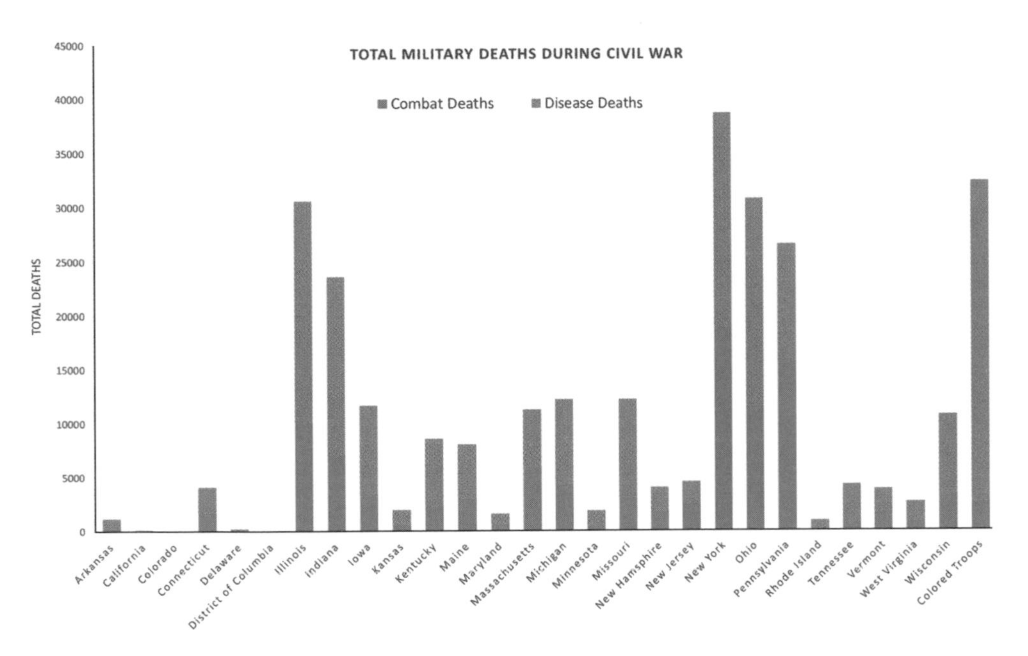

Figure 11.3
It's easy to see that deaths due to disease (the red bars on top) significantly outnumber deaths due to actual fighting.
Credit: Daniel M. Russell. Redrawn chart based on CivilWar.org's data set, https://www.battlefields.org/learn/articles/civil-war-casualties

Naturally, this made me curious about what "rose catarrh" meant back in those latter days of the 1800s. My query was:

["rose catarrh" 19th century]

I was happy to see not just some articles on it but also images that lead to newspaper stories from the period (figure 11.4).

With these hits, it's easy to quickly understand that rose catarrh is just what we now call hay fever—that is, an allergic reaction to pollen, dust, and so on. In the late 1800s, it was one of those things that caused all kinds of advertisements, each peddling patent medicine cures. Here's one from the *Los Angeles Daily Herald*, March 31, 1887, 6 (figure 11.5).[6]

So rose catarrh won't kill you; you won't die from it, you'll just feel crummy. And of course it's a phrase that's fallen out of the language along with flux and quickstep.

Figure 11.4
Searching for the phrase "rose catarrh" and the context term "19th century" gives a set of results that tells us that it's just a variety of hay fever. The images of "rose catarrh" on the page link directly to news stories (and advertising) from that period. Credit: Google and the Google logo are registered trademarks of Google Inc., used with permission

Reading widely is a dangerous thing to do. As I was reading about the Civil War era, as I was reading through newspapers of the time about rose catarrh, I was struck by the number of times that I saw stories and advertisements for the amazing curative and restorative power of hot springs. If the war didn't kill you, a dip in the hot springs would surely make you better. But then I was led to this research question, which naturally follows from the first one.

Research Question 2: *These days, it's popular to go to a spa that features natural hot springs, such as those at Big Sur (California), Wiesbaden (Germany), or Bath (England). But if I'm searching for such a spa to visit in 1890s' America just after the Civil War, what search terms should I use?*[7]

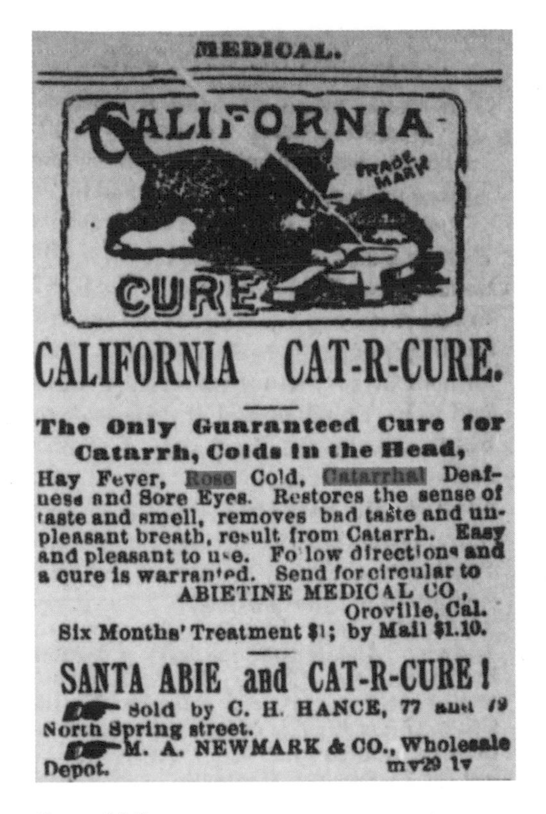

Figure 11.5
This advertisement for the California Cat-R-Cure (get the joke?) claims to cure catarrh generally and "rose cold" in particular.
Credit: Library of Congress, *Los Angeles Daily Herald*, March 31, 1887, 6, image 6; image provided by University of California at Riverside

Again, the problem here is to find terms that were used *during this time* to describe a spa with hot springs.

As we did above, you can get started by doing standard synonym queries and poking around in Wikipedia articles, looking for alternative phrases.

But I decided to go to one of the contemporary sources: archival newspapers. I found the Library of Congress's Chronicling America newspaper collection by doing a search for the following:[8]

[online archival newspapers]

Then, once there, I limited the dates of my search to 1890–1899, and searched for "Saratoga Springs" (a place that I already knew as a famous

Figure: 11.6
A sample search for "Saratoga Springs" on the Library of Congress newspaper archive website. Each of the red boxes shows a hit for "Saratoga" or "Springs."
Credit: Library of Congress; Copyright Library of Congress

natural springs resort popular during those days). My idea was to search for newspaper accounts of the place and see how the *articles* referred to it.

My query gave me a page that looked like the following (figure 11.6).

All the red boxes you see on the page are hits for the string "Saratoga Springs."

Within a couple of minutes, I found those springs referred to as: *sanitarium*, *tonic spring*, *sulphur springs*, *soda springs*, and commonly, just plain-old *hot springs*.

I'm sure that if I kept reading, I'd find other ways to refer to these kinds of therapeutic resorts with springs. But this is a great list, and it comes directly from the writing of the times.

Interestingly, the best way to pick up on these alternative descriptions was to read through the advertisements. Often the language is colorful and the claims are extravagant yet completely fun to read! Here's one from a place at the edge of today's Silicon Valley.

Figure 11.7
Ad for Gilroy Hot Springs, California, in the *San Francisco Call*, July 9, 1895, 11
Credit: Roop & Son, *San Francisco Call,* August 8, 1895. Digitized by UC Riverside

I too want to relax at a place where the "Ogre Malaria Never Lifts His Ghastly Head and Where the Waters of Healing Pour Freely from Nature's Own Fountain."[9] And now I know that if I want to look for all the expressions of what we think of today as "hot springs," I should also search for "tonic spring," "soda spring," "sulphur spring," and so on.

That is, I'd do a search like this:

> [19th century "hot springs" OR "tonic springs" OR "soda springs" OR "sulphur springs"]

Such a search would get the widest-possible range of results for the warm, healing, sometimes-smelly waters that people of the day would enjoy.

Research Question 3: *While reading about the Civil War time period, my attention wavered for a few moments as I read about the binoculars of the period, thereby leading me to read about the life of John Dollond (the inventor of the achromatic lens, for which you should be grateful), and I learned that he died of a stroke. But oddly, I can't find period accounts with "stroke" as a search term. Is this another term that has changed since the mid-nineteenth century?*

What search term should I use instead to find an eighteenth-century death by stroke?

I spent quite a bit of time searching the newspapers and books of the time for "John Dollond" and "stroke," but not getting much of any useful results.

Now we know that searching directly for *synonyms* can often give us lots of options, but then we need to verify that the term was used during that time period.

My first query was:

[stroke synonym]

That led me to the only term I hadn't heard of before: *apoplexy*. (Well to be honest, I'd heard of this word before, but I wasn't really sure what it meant!)

I did a quick define search:

[define apoplexy]

It confirmed that it means a "stroke."

How do we confirm that this was being used in the 1700s? The Ngram's database only goes back to 1800, and many of the newspaper archives are limited to post-1800 as well.

But Books.Google.com goes WAAY back! By searching in Google Books for "apoplexy" and limiting the time range to 1700–1799, we see the following (figure 11.8).

I scanned down a bit to find a book that has a readable (to my modern eyes) explanation of apoplexy and found it in the 1793 book by Samuel Auguste David Tissot: *Advice to People in General, with Respect to Their Health. Translated from the French. ... To Which Are Added, by the Author, Two New Chapters; One upon Inoculation, the Other upon Lingering Distempers. ... The Sixth Edition, Corrected and Improved.*[10]

In this book, we find the following definition (figure 11.9).

I like the image here of "inflammable blood, and that in a large quantity ..." But you get the idea. Apoplexy is the word we seek when we need to search for people who died of a stroke in the eighteenth and nineteenth centuries.

In particular, using this new (old!) synonym for "stroke," I searched for:

[John Dolland apoplexy]

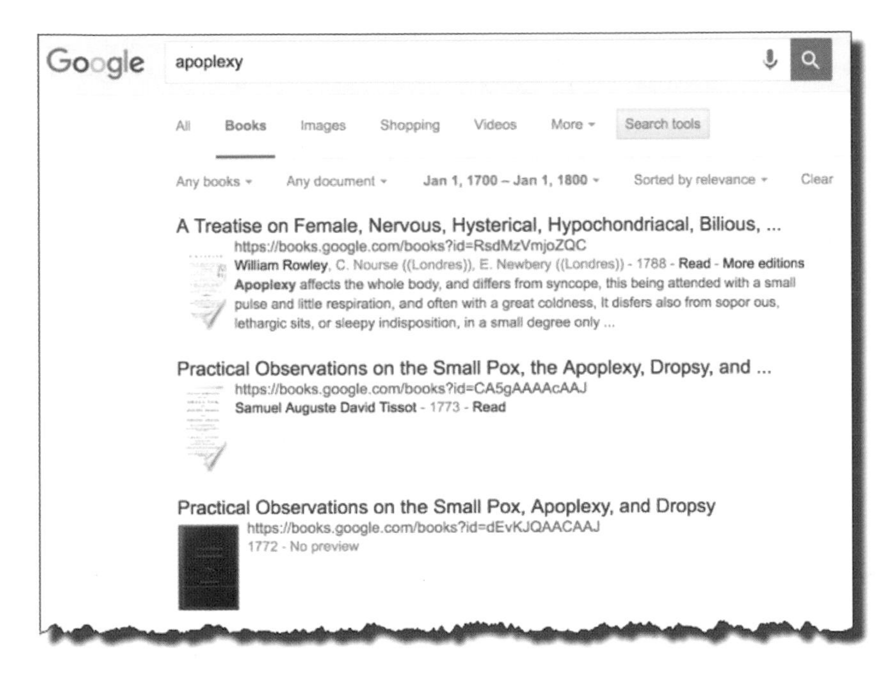

Figure 11.8
A search for "apoplexy" in Google Books with the date range restricted to the eighteenth century shows a number of books that you can read. If you check some of these references, it's clear that "apoplexy" means "stroke" in a medical sense.
Credit: Google and the Google logo are registered trademarks of Google Inc., used with permission

I quickly found an account of his death in the *Gentleman's Magazine* (1820), not long after he was appointed to be a Fellow of the Royal Society *and* optician to the king. As the article points out, he "did not live to enjoy his honours long, as he died of apoplexy, Nov. 30, in the same year."[11]

But you see the point. To find this account of his death, we had to use language appropriate to the time. In this case, we had to search for *apoplexy* rather than *stroke*.

If you want to search for terms that are even older than the nineteenth century, you'll want to know about a *historical thesaurus*, such as the *Historical Thesaurus of English* (for English words over the past several centuries).[12] This is the resource you'll want to use if you're doing research in a time period that's different from your own. (And it's an alternate way to do this particular search.)

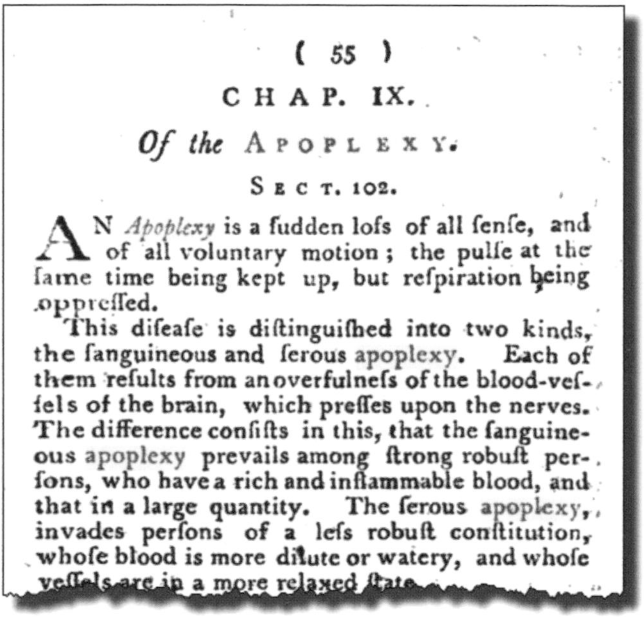

(55)

CHAP. IX.

Of the APOPLEXY.

SECT. 102.

AN *Apoplexy* is a fudden lofs of all fenfe, and of all voluntary motion; the pulfe at the fame time being kept up, but refpiration being oppreffed.

This difeafe is diftinguifhed into two kinds, the fanguineous and ferous apoplexy. Each of them refults from an overfulnefs of the blood-veffels of the brain, which preffes upon the nerves. The difference confifts in this, that the fanguineous apoplexy prevails among ftrong robuft perfons, who have a rich and inflammable blood, and that in a large quantity. The ferous apoplexy, invades perfons of a lefs robuft conftitution, whofe blood is more dilute or watery, and whofe veffels are in a more relaxed ftate.

Figure 11.9

Page 55 of the above-mentioned book. Recall that the letter that looks like an *f* is in fact a long *S*. Thus, "fudden" is "sudden," and "ferrous apoplexy" is "serous apoplexy," where "serous" means "of serum"—that is, an apoplexy that was believed to be caused by an excess of a watery serum in the brain.

Credit: Google; Copyright Google Books

The *Historical Thesaurus of English* has all the words from Old English days until now—in both obsolete and current definitions—all extracted from the *Oxford English Dictionary*, and organized by their meanings and dates of use. This way, each word shows up in its own era, demonstrating how even simple words have changed.

For instance, if you want to look for a disease symptom like "fever," you'll need to use different words depending on the time period that you're interested in. Using the *Historical Thesaurus of English*, you'll learn that "fever" would be written as *feferadl* in the tenth century, *febricitation* in the sixteenth century, *pyrexia* in the eighteenth, and *febrility* in the nineteenth.

In the end, language shifts happen all the time—and relatively quickly. As we found out, you can die of apoplexy, but not of rose catarrh. On the other hand, if you do have rose catarrh, you might well have a feferadl.

Research Lessons

1. *Search for synonyms, but check that the synonyms you find are appropriate to the time period.* As we saw, the obvious search for synonyms will often offer up old-fashioned and even archaic terms for an idea that you seek. But you have to check that this term is the right one for the time. You can search for synonyms for terms using a context term like "19th century," but remember that newspapers or books of the period can be incredibly good resources. And if you go far enough back, check a historical thesaurus for specific terms in that specific time period.
2. *You might find multiple terms!* It shouldn't be a surprise, but people frequently have different ways to refer to a common concept. (If you think about it for a second, you can probably call to mind a few different ways to refer to diarrhea in your own language. It shouldn't surprise you that this was true in the past as well.)
3. *A great way to find old terminology is to check out old newspapers and books.* The simplest way to do this is via online archival newspapers (e.g., those at the Library of Congress newspaper collection, Chronicling America) or in a books database that has a collection from the deep past (such as Books.Google.com or the Hathi Trust).[13]
4. *If you go far enough back, sometimes the characters themselves are different.* Text that looks like a "fudden lofs" is actually a "sudden loss"—especially true in books from the eighteenth century. In Google Books, however, the long *S* character is recognized as an actual *s*, so when you search within the text of the book, just search for "sudden loss" and you'll find the correct text (even though it might not look like it).

Try This Yourself

Language changes over time, often in profound ways. It's important to know this when you're searching for something archival or historical.

If you want to see for yourself, try searching for information in a topic area that you know well—and then carry that search back in time for sources that discuss your favorite things, but from a century earlier.

Say that you're interested in football (the US kind). How has the language of that sport changed over time? What's a "momentum mass play," "wingback," or "rush line"? Or in baseball, what was a "Baltimore chop" or "dirt dipper"?

Likewise, you could look at the changes in language for a technology over time. Train locomotives have changed considerably over time. What's "induced drafting" or a "thermic syphon"? Do those devices have equivalents in today's locomotives, or have they, like buggy whips, vanished into the dark recesses of language?

As you look backward in time, consider how you might find collections of terms that would be useful to you. How do you find old-fashioned, out-of-date, or archaic terms in a particular field?

12 What's That Wreck Just Offshore? How to Find Archival Imagery and Use Metadata from Photographs

A strange pile of wreckage just offshore, map, strange story of wrecks, dancing pavilion, and possible arson. How can we solve a cold case through online research? Learn how to use online maps and archival news sources.

On a Saturday, I went for a run on a beautiful summer day along the shoreline of the Carquinez Straits, where the Sacramento River flows serenely into San Francisco Bay before heading out to the Pacific Ocean. As I cruised the southern shore of the straits, I ran past a big pile of iron and steel sticking up out of the water just a few yards offshore in fairly shallow water (figure 12.1).

Being a curious sort of fellow, I took a picture with my phone and kept going on my run, knowing that later I could figure out what it was all about.

But seeing this wreck left me with a few questions.

Research Question 1: *What's the story of this wreck? What was it? Did it have a name? Why is it here?*

First I knew it was a cell phone photo, so there would be EXIF metadata attached to it. It's a simple process to extract the EXIF data from the photo[A]. I also knew that the EXIF metadata would have a fairly accurate latitude and longitude of the wreck's location. It turns out that when I took the picture, I was standing at 38.054242, –122.20228.

I started by going to Google Maps to get a look around, just to see what things might be nearby that I could use as search terms. Here's what I saw when I put that latitude/longitude into the Google Maps search box (figure 12.2).

But that's too wide a view of that location on the map to be useful. I thought I should zoom in a bit to see more detail.

Figure 12.1
Seeing this wreckage just a few yards offshore made me curious. Why is there a shipwreck in the Carquinez Straits, not far from San Francisco Bay?
Credit: Daniel M. Russell

And what do you know? There's the answer right there on the map! At exactly the place where my "hunk of junk" lay, the map has a label in the water marked as "Garden City Wreckage" (figure 12.3).

Longtime Google Maps users know that there are layers of information that show up at different zoom levels. Unfortunately you can't turn them on/off individually but instead have to manually zoom in and out until the layer you want appears. In this case, I got lucky that it had shown up. (If it hadn't, I would have used other clues shown here such as "Eckley Pier," "Bull Valley," or the nearby town of "Port Costa" as place-names. I'd have found it that way.)

Yet as often happens, someone else had already marked this particular spectacular junk pile. I switched to Google Earth view and zoomed way in to see if it was what I thought it was. Here's the satellite image (and it's clearly the same, although at low tide). (See figure 12.4.)

Figure 12.2
Entering the latitude/longitude into Google Maps shows that place.
Credit: Map data © Google

Figure 12.3
By zooming in a bit, more details are revealed—including the location of the wreckage and its name.
Credit: Map data © Google

Figure 12.4
Credit: Map data © 2014 Google

Now that I know it's the "Garden City" (and probably a boat of some kind), I had to figure out the story.

I did a query for:

["Garden City" wreckage]

The Google News archives returns a number of hits, many of which tell the story of how the good ship *Garden City* was built in 1879 as a side-wheel ferry for the South Pacific Coast Railroad to move train cars to San Francisco.[1]

Although the *Garden City* was built with narrow-gauge track on the main deck to carry freight cars to San Francisco, it could also carry passengers and cars as a relief ferry. Southern Pacific used the *Garden City* as a relief boat for its auto ferry run on the old "creek route." The *Garden City* stayed on the creek route as a passenger ferry when auto ferry service was shifted to the Oakland pier.

The ferry ran from Alameda to San Francisco until 1929, when it was brought to the piers at the town of Eckley as a fishing resort and bit of a

Figure 12.5
In this shot from the water, you can clearly see the boilers and engine remains from the *Garden City* ferry.
Copyright Jafafa Hots, https://www.flickr.com/photos/jafafahots/

dance hall. Unfortunately that was its last assignment before becoming a picturesque wreck.

Of course, now I'm curious about what this wreck looked like *before* it became an odd pile of metal sticking out of the water near San Francisco.

Research Question 2: *Can I find a picture of the* Garden City *ferry that was taken BEFORE it turned into a watery ruin?*

This task wasn't that hard. I already knew what it was and its name:

["Garden City" ferry]

Once you've done that search, you can switch your search mode to look in Google Images; you'll find that there are quite a few images of the *Garden City* before it met its untimely end.

Figure 12.6
The *Garden City* ferry pulls up at the Oakland Municipal Wharf, at the foot of Franklin Street (ca. 1900).
Credit: Courtesy of Oakland Public Library, Oakland History Room and Maps Division

For instance, here's an image I found of the *Garden City* steaming into the wharf at Oakland, not that far from where it ended up (around 1900).

As you can see, in its heyday, the *Garden City* ferry was in a world of horses, wagons, and early cars. It was truly from another era.

Another great resource to know about when searching for geographically based information is Wikimapia. It's a privately owned, open-content, collaborative mapping project that's not actually part of the Wikipedia world, but shares its philosophy in encouraging people to share their knowledge about pieces of the world that are then layered into its maps.

Armed with the latitude/longitude, you can look up Wikimapia's annotation about the *Garden City*.[2]

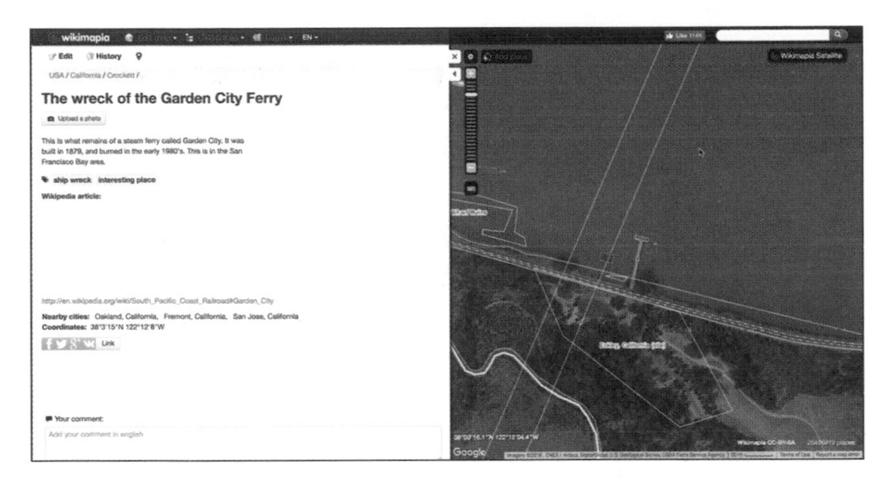

Figure 12.7
The Wikimapia entry for the *Garden City* ferry wreck gives the built-on date and an estimate about when it burned.
Credit: Wikimapia; Map data © 2018 Google, CNES/Airbus, DigitalGlobe, US Geological Survey, US Department of Agriculture's Farm Service Agency

From the Wikimapia site, I again saw that the wreckage was there because the docked ferry had burned at some time in the 1980s. Now how did THAT happen? That led to my next query:

["Garden City" ferry burn]

With this query I found the PrivateNaturalist website, which tells its own story of searching for the *Garden City*, noting, "*This rusting heap is the boilers and paddle-wheel hub of the old Southern Pacific Ferry Boat 'Garden City.' ... [I]t burned to the waterline in the fire of 1983.*"[3]

There's another version of that same clue. "The fire of 1983."

This kind of thing is just crying out for further investigation. It burned in 1983, in the straits, not far from town, so there must be some kind of contemporary account. How do I find that?

Research Question 3: *Is there a newspaper article about how this wreck came to be?*

At this point, I want to read something that was written at the time of the *Garden City* ferry, and in particular, about how it became a picturesque

Lodi News-Sentinel - Aug 9, 1983 Browse this newspaper » Browse all newspapers »

Port Costa fire is contained by firefighters

By United Press International

Eight arson-ignited fires that fused together to destroy several homes, outbuildings and a century-old ferryboat sizzled through tinder-dry brush around historic Port Costa, Calif., Monday.

The late Sunday blaze, which blackened 400 acres 30 miles east of San Francisco on the edge of San Pablo Bay, was contained during the night, but firefighters said "hot spots" flared up, keeping them from gaining full control.

"We're concentrating on the 'hot spots' right now," fire department dispatcher Donald Tacconi said. "The valleys have many eucalyptus groves which are burning."

In Oregon, forestry officials were warily bracing for "dry lightning" storms predicted for late Monday. Lightning bolts have already touched off dozens of fires in Oregon and neighboring Idaho, blackening more than 200,000 acres during the past week.

A blaze near Mountain Home, Idaho, spread to 23,000 acres Monday before it was declared contained. However, a new range fire was reported near Stone, 60 miles south of Pocatello near the Idaho-Utah border.

Temperatures soared toward the 100-degree mark in the West in the continuation of a nationwide heat wave.

Figure 12.8
One of the stories about the burning of the hills, and the *Garden City* ferry, near Port Costa.
Credit: *Lodi News-Sentinel*

wreck just offshore. Now it's time to turn to Google's Newspaper Archives.

To use it, I went to News.Google.com/newspapers and redid my query there. This search leads to several articles in the news archives, all saying more or less the same thing. Fires, started by arsonists, destroyed large parts of the area and spread onto the old wooden ferry, burning it to the waterline next to the Eckley pier on Sunday, August 7, 1983.

The arsonist aspect of the 1983 fire was of interest to me so I searched papers to see if the person was caught. The fire chief's son from the nearby town of Rodeo was responsible and pleaded guilty. He was sent for psychological testing. His father was to retire the next month.[4]

By doing a bit more digging in the newspaper archives (searching for news articles with the search query [arson ferry Port Costa 1983]), I found a news article from 1985 that indicates a local father turned in his son for the arson after admitting that he was the one responsible.[5] That's a difficult thing for any father to admit, perhaps especially one who is the local fire chief.

Now I'm curious about the history of the ferry and the role that it played in the transportation around San Francisco Bay before the bridges were built. That made me realize that I should also check Google Books for any mention of the *Garden City* ferry and found a few with this query:

["Garden City" ferry]

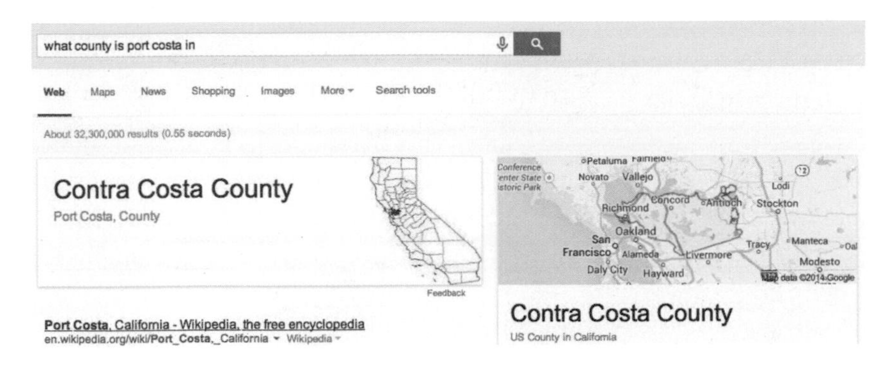

Figure 12.9
Port Costa is in Contra Costa County.
Credit: Google and the Google logo are registered trademarks of Google Inc., used with permission

It was a quick click to find *Ferries of San Francisco Bay* and then search in this book for "Garden Ferry."[6] There are four great photos of this formerly grand ferryboat on pages 27 and 28.

Another excellent book in the same search results page is *Port Costa,* which also tells the history of the ferry, with many images over the years, and with a detail saying that it was the "largest dance floor in Contra Costa county" in its day.[7] There's even a Christmas card from its time when the ferry was a fishing resort.

When you're searching for historical information like this, you should remember that there are also often countywide historical societies that collect information (and photos). To extend my reach, I thought I'd look for a possible countywide library or archive that might have additional information. Since Port Costa is the closest town to the wreck of any size (and it's pretty small), I did the obvious query:

[what county is Port Costa in]

Note that this gives us a name and a map (figure 12.9). Now that I know the county name, I can look for:

[Contra Costa historical society]

In this way, I find the Contra Costa Historical Society.[8] It has an old-school site, but by using the built-in search tool, I was able to find many more photos of the ferry and get confirmation from a different (and local source) about the history of the ferry as well as how it came to be mired in the muck near the Eckley pier.

Another useful resource to remember is that many local city and county libraries have stashes of archival materials. The downside of these local archives is that they're often NOT indexed or searchable online. This is another case when you just have to go there to get the information; it's simply not online.

But where is *there*, exactly? How can you locate all the libraries in a region?

When I search for local libraries and archives, I generally use Google Maps. In this case, I did a search for:

[libraries in Contra Costa county]

I found all the county branches and city libraries. If you click on one of the drop pins, you'll get more information about that library, including a link

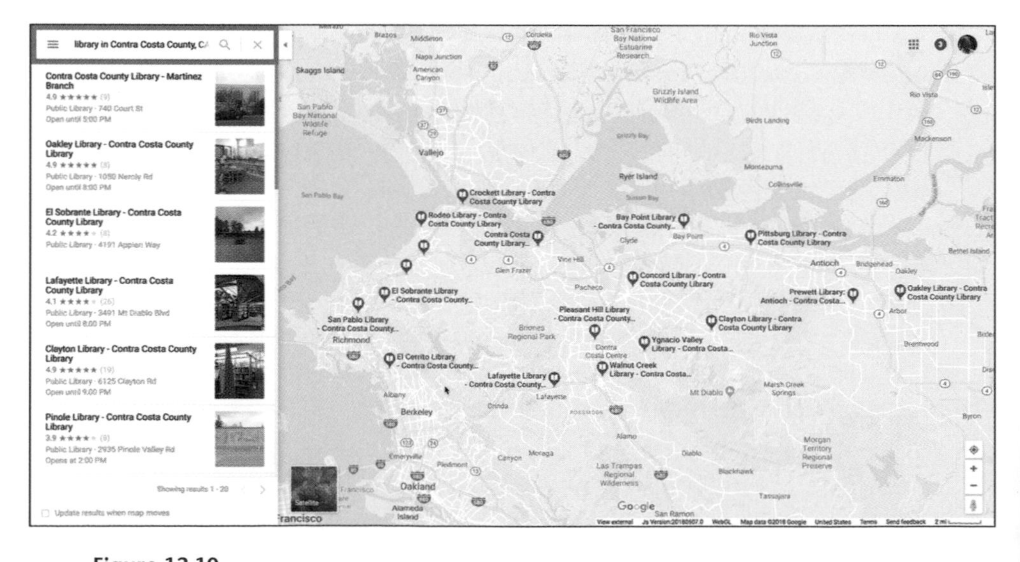

Figure 12.10
A search in Google Maps for [library in Contra Costa county] shows all the public libraries in the county. Doing a similar query with "historical society" or "museum" can find even more sites with useful archival materials.
Credit: Google and the Google logo are registered trademarks of Google Inc., used with permission

to its website, but also the phone number. Frequently, to figure out what they've got, you actually have to call and do a bit of off-line research. But at least you now know where all the potential archives are.

Research Lessons

1. *Zooming in on a map can sometimes reveal information that you can't see at all levels.* In this case, it basically gave me the answer instantly, but you have to know to zoom all the way in (and check each level—most easily done by using the + and – controls).
2. *Switching the satellite view can reveal new information.* In this case, we can see that the wreckage is accessible at low tide. (I'm tempted to go back during an extreme low tide just for photographic purposes!)
3. *Using local place-names can help focus the search and find new information.* In this case, Eckley is the name of the pier, but it WAS the name of the village where the *Garden City* was finally moored. Likewise, Port Costa is the closest "big" town (although nowadays, it has a population of around two hundred people), but it's a longer-lasting place than Eckley, which no longer has any town left. BOTH place-names, though, help to find news articles and books with local information.
4. *Local historical societies and libraries often have great local historical information.* Here I used the "look for a county historical society" trick to locate a nearby cache of historical photos. Keep that in mind when you're searching for this kind of information. Often, historical societies' archival content is off-line or difficult to search. A phone call and real-life visit will frequently give you great results through the help of friendly librarians and archivists.

How to Do It

A. Extracting the EXIF data from the photo. Almost all digital cameras capture some metadata and store it with the image. For JPG and TIFF files, this is called EXIF metadata. (For details, see the Wikipedia article on EXIF.)[9] Settings like shutter speed, ISO speed, white balance, aperture setting, focal length, and location in latitude and longitude are all recorded and stored when the photo is taken.

For OUR purposes, it's enough to know that EXIF almost always has all the information about the picture, and if it was taken on a smartphone, it will have the GPS data as well.

There are lots of tools for inspecting the metadata of an image. You can get some of the metadata by getting "more information" when you use the preview app to look at an image on the Mac. Viewing EXIF data in Windows is simple as well; just right-click on the photo and select "Properties." You can also find lots of tools for reading the metadata. I did a search for [image metadata tool] and found several. But getting the EXIF data is fairly straightforward. Seek and ye shall find.

Try This Yourself

The world is literally covered in mysteries that you can uncover with a bit of research. There are certainly some near you, but if you'd like to spend a little time researching some fascinating mysteries, here's one to try out on your own.

San Francisco has many venues for entertainment of all kinds, but perhaps one of the more interesting is this ruin that's on the western shore of San Francisco near Seal Rocks. Using your online research skills, can you figure out what this set of ruins is and its history (figure 12.11)?

Caution: this particular place is so rich in history that you might end up spending far more time reading about it than you'd imagine. Once you determine what it is and read about it, look around for some images of the place before it fell into ruin.

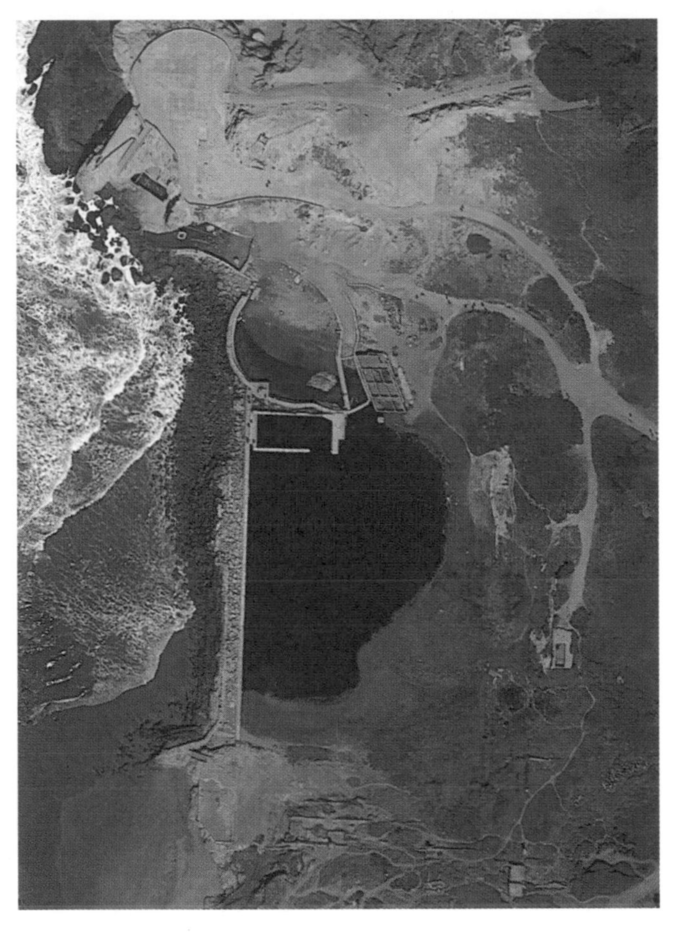

Figure 12.11
What happened to make this intriguing ruin on the western edge of San Francisco? This used to be something spectacular. It's at latitude 37.780148, –122.513849, and that long wall running north to south is 225 feet long (or 68.5 meters).
Credit: Google Earth

13 Do Flies Have the Pattern of a Spider on Their Wings? How to Check the Credibility of a Resource You've Found

Some fruit flies have a remarkable pattern on their wings. Is it the image of a spider? Maybe an ant? And how is such a thing possible?

As often happens while reading stories online, I find a slightly incredible story and wonder if it was true. This probably happens to you too; here's what I do about it to satisfy my curiosity.

In this case, the slightly incredible story started with a tweet about how a small fly apparently had images of ants on its wings. The comment stream on the tweet wondered if it was an evolutionary master stroke, a divinely ordained pattern, or just randomness at work. This image was such a remarkable thing that I just had to go check it out and see for myself. This in turn led me to do a small investigation into flies, wings, spiders, and evolution. Here's the story, which I hope will shed some light on the process I use for evaluating incredible stories, trying to find out if they're believable or not. Here's the original tweet by @ziyatong in November, 2013 (figure 13.1).

This tweet was startling, making me curious, and that led me to start thinking about what it was that caused this reaction. Is there a way to distill a research question here? My initial attempt included two research questions.

Research Question 1: *Is this fly for real, or is this all some kind of strange hoax?*

Research Question 2: *Does this fly actually have the image of ants on its wings? (That is, are they supposed to look like ants?)*

Figure 13.1
This is a remarkable image of a set of fly wings with strange markings that seem ant-like—on a fly with extraordinarily beautiful eyes.
Credit: Peter Roosenschoon, photographer; Ziya Tong, Twitter poster, https://twitter.com/ziyatong/status/397348948857196544

I started this research process with just the Latin name of the fly that the questioner wrote in the beginning of her tweet. Is that the right name for this fly?

[Goniurellia tridens]

It turns out that if you know it, the scientific name is often a really great place to start a search. It's almost always an unambiguous name for an animal or plant, perfect for use as a search query.[1]

This search finds a lot of online content, including a blog post by evolutionary biologist Jerry Coyne titled "Why Evolution Is True" that discusses this fly with the curious wings.[2]

The buzz about this image is that it's strange and remarkable. How could a fly have evolved to have TWO images of other insects on its wings?

Just to be clear, I'm not making a statement here about evolution or even flies. (At least not yet.) This is a great example of things that cause you—when you read something or see an odd picture—to pull up and say, "Hey, wait a second. Let me look into that a bit more."

You know what it's like. You're reading along and then something grabs your attention—in this case, the fly with images of other insects on its wings. My first reaction was that this is pretty amazing. But then my "critical response" kicked in. And this is probably the biggest lesson here:

> *Whenever something strikes you as especially remarkable, wacky, or surprising—whenever you say "huh!"—you probably ought to check out the story.*

This is really the basis of all research. As Isaac Asimov purportedly said, "The most exciting phrase to hear in science, the one that heralds new discoveries, is not 'Eureka' but '*That's funny.*'"[3]

And that's the case here. If something stands out from the background, it's because you've seen something that violates your expectations. It's "funny." Perhaps it's a stroke of color in an oil painting that you didn't expect or the strange appearance of an insect on the wing.

So when I saw this, my Spidey sense started tingling, and I (being a curious sort of fellow) did a little digging.

I first read the post on that blog carefully. I noted the scientific name of the fly (*Goniurellia tridens*), the photographer (Peter Roosenschoon), where the photo was taken (Dubai Desert Conservation Reserve), when (January 2010), and so on. I then followed the links back to the Dot Earth column in the *New York Times* and read that.[4]

Interestingly, since I first read the *New York Times* article, an update has been made. At the top, the article has a few new lines of text that link to another blog, *Biodiversity in Focus*.[5] This blog is written by an entomology graduate student, Morgan D. Jackson, and as is the case with many graduate students, they're often the people most informed about the latest insights in their field. In this instance, his blog article on this topic, "Ants, Spider, or

Wishful Thinking," is excellent, and offers a learned background in a story that's well worth reading.[6]

But already, just in a few clicks, I've discovered that there's a range of opinions about this tiny fly. I found the original article, the original article it cited, and then the modified blog post. What's going on here? What should I be thinking about this?

This led me to the most basic questions about the authors.

Research Question 3: *Who are these people? Should I believe their images and posts? What are their backgrounds?*

The original tweet poster says that she got the image from Roosenschoon. It's not hard to find evidence that Roosenschoon really did take this photo, and that he really does work at the Dubai Desert Conservation Reserve; since his name is uncommon, a search like [Roosenschoon fly wings] finds his photos. Likewise, searches for his photos show us that he takes pictures of desert animals.

This kind of quick verification check is something I do nearly constantly. When I run across a name I don't know, and if what that person is saying is important to the story, I'll do a quick search for their name and the topic to see if they actually work in this field, or have written something else on this topic.

So next I checked out the author of the *New York Times* blog post (science writer Andrew C. Revkin) and Jackson (an entomology graduate student who studies flies). They both have writings, images, and publications that you can easily discover. They are who they say they are. They have skills and background knowledge that's relevant to say smart things about the wing patterns of flies.

Now I can reframe my first question a little more precisely.

Research Question 4: *Are those images of insects on the wings of this fly?*

They certainly look insect-like, but as Jackson points out in his blog post, it might be just an accident of wing coloration patterns that look like insects to humans. You could really think of this as a Rorschach inkblot test; it could be a random pattern that looks like images of bugs on the wings. So I'd say no—at this point, I'm not convinced that they're really *supposed* to be depicting bugs. But it's easy to find many pictures of this fly, *Goniurellia tridens*, and see that the image hasn't been edited. (Keep reading, though; sometimes things in the image that aren't edited could be significant.)

Research Question 5: *WHY do you believe whatever it is you believe about this?*

I'm certainly struck by the depth and careful analysis of Jackson's commentary on this. Yet being skeptical myself, I went and looked up some original source material.

I know that animals within a family often share many characteristics. So I did a search on the family name of the fly, *Tephritidae*, in Google Books and found more than a few entomology books that mention this family name. Here's one I discovered that was pretty interesting in an insectoid way: *Fruit Fly Genera South of the United States (Diptera: Tephritidae)* by Richard Herbert Foote.[7] That's an unwieldy title for a book, but a quick look inside tells me that this is a deeply serious work by an academic author *for* academic readers.

Yes, this is a book about related flies from the New World while the fly was found in the Old World, but the family and even genus is the same, and so I poked around a bit in this book and found the wings from closely related fruit flies. I just searched in the text of the book for [wings] (by using the "**search in this book**"[A] function) and found a perfect illustration that compares different wing patterns from a variety of fruit flies (figure 13.2).

And just for grins, I extracted the wing image from the picture above and put it side by side with the wing of the fly in the top picture. I converted it to gray scale so it would look pretty similar to these other pictures and then erased a bunch of the peripheral clutter. Here's the end result, with the other image side by side for comparison (figure 13.3).

As you can see, these look pretty similar; not quite exactly the same, yet fairly close. I'm not sure how variable these wing patterns are, but this definitely seems like it's within the normal range of variation. Now that I've noticed this variation, it inevitably leads to my next research question.

Research Question 6: *How variable are these kinds of fly wing patterns?*

To be clear, what I'm trying to figure out is if all these fruit flies have patterns on their wings, and if so, how different would the patterns be from each other? If they all have patterns, and they all vary quite a bit, this

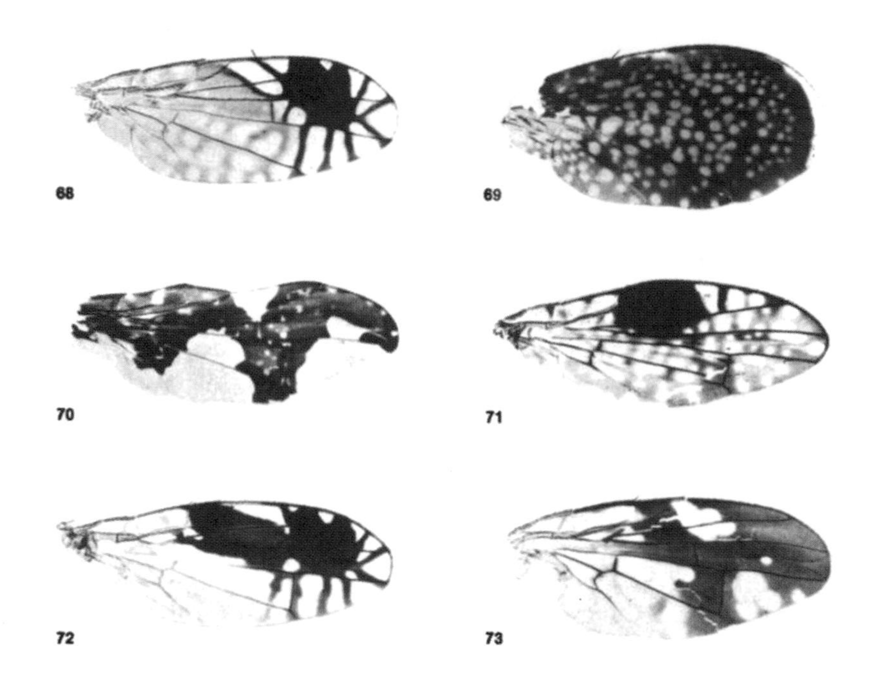

FIGURES 66–73.—Wings of Tephritidae: 66, *Euaresta reticulata* (Hendel); 67, *Euarestoides acutangulus* (Thomson); 68, *Euarestoides* sp.; 69, *Eutreta hespera* (Banks); 70, *Gerrhoceras* sp.; 71, *Gonioxyna* sp.; 72, *Goniurellia* sp.; 73, *Gymnocarena* sp.

Figure 13.2
Side-by-side *Tephritidae* fly wings of different species, including *tridens*.
Credit: Herbert Foote

would suggest that the wing patterns are built in and not necessarily an image of another insect.

My query to figure this out was:

[Goniurellia wing patterns]

And I clicked on the images that were offered. In there, I found that not only do many of the species variations of *Goniurellia* have patterns on their wings but the patterns also vary quite a bit (figure 13.4)! Along the way, I serendipitously discovered that there are several other kinds of flies (including flies that live in the United States and United Kingdom) that are called "picture wing flies" simply because they have transparent wings, with "pictures" on them, rather like our *Goniurellia*.

Figure 13.3

Side-by-side comparison of the original fly wing (by Roosenschoon) and the fly wing from the previous figure—the upper-left wing image. You can see how one might be a small variation on the other.

Credit: Daniel M. Russell

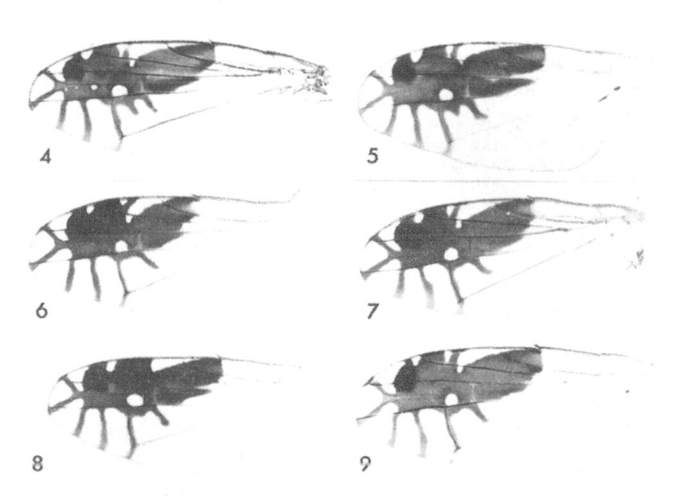

Figs 4–9. Wings of *Goniurellia* species. 4. *G. tridens* (Hendel). 5. *G. longicauda* spec. nov. 6. *G. munroi* spec. nov. 7. *G. omissa* spec. nov. 8. *G. persignata* spec. nov. 9. *G. spinifera* spec. nov.

Goniurellia wing patterns, from Freidberg, 1980
(Journal of the Entomological Society of Southern Africa 43(2):257-274)

Figure 13.4

The variability of *Goniurellia* wings across different species.

Credit: Amnon Freidberg, "A Revision of the Genus *Goniurellia Handel* (*Diptera: Tephritidae*)," *Journal of the Entomological Society of Southern Africa* 43, no. 2 (1980): 262.

Now that I've learned about picture wing flies, the inevitable next query is:

[picture wing flies]

Here I learn that picture wing flies are nearly always fruit flies (like our *Goniurellia*), and that there is a huge variety in the different patterns (pictures) that can be seen on the wing.

More to the point, one of the results from this search is a paper in the science journal *Nature* that tells us the images on a fruit fly wing are determined by variations within a single gene. All these variations in the wing pattern involve the repeated use of coloration that is already built in to the wing (figure 13.5).

It IS remarkable to see these patterns that look like insects on the wings of other insects, but perhaps more in the same way that a cloth can be said

Figure 13.5
Fruit fly wings from across the world show a huge number of pattern variations, all controlled by variants on a single gene.
Credit: Nicolas Gompel and Benjamin Prud'homme, https://www.sciencedaily.com/releases/2005/02/050205080226.htm

to have an image of Christ on it. (I'm not making this up; this happens often enough that there's even a word for such things: *veronicas*.)

I think we've established that these wing patterns are real, and that there are an awful lot of patterns on fly wings all around the world. But are they *supposed* to look like ants or spiders?

Perhaps more to the point, is this remarkable fly even doing something that's normal? In his blog post, entomologist Jackson points out that

> if you look closely at the photo you can see that the middle and hind legs are actually curled up under the body and the fore legs are not resting in a natural position at all—they look more like a ballerina *en pointe*, with the last leg segments curled and with the tops of the "feet" (the tarsomeres) resting on the surface of the substrate—which leads me to believe the fly in the photo may be dead, or at least heavily compromised, and not actively displaying its wings at all![8]

In any case, in a dispute among experts, as is the situation here, the entomologist is going to win. This is his area of expertise, his argument is well reasoned, and he gives a good bit of background information about mimicry in other animals and goes on to discuss the role of functional mimicry (as opposed to accidental imitation) that we humans perceive as being a mimic. He also brings up a significant question about something that I hadn't even noticed: Is this fly even alive? Or is it in some strange altered state? The fact that an entomologist is questioning this (who not only knows what the fly is supposed to look like but also knows the specific word for the fly leg segment that makes up the foot of the fly) leads me to think that he knows what he's talking about. Sometimes estimating the credibility of an author on a given topic rests on estimating how much knowledge they demonstrate about the topic area. Knowing about the normal standing posture of a fruit fly suggests that this person knows their stuff.

But there's more to this ability to tell when a source is credible or not. Part of understanding what it means to find a credible resource relies on knowing something about that source as a producer of reliable content.

The fact that the *New York Times* posts updates to its science columns with information that amplifies and somewhat contradicts its original posting, AND that it didn't modify the original article to make it seem like the authors knew what they were doing all along, are good signs. They suggest to me that the *New York Times* science writers are being truthful, honest, and open in their disclosures about what they knew, and when they learned it. Another hallmark of credible writing is the willingness to admit to errors in earlier work. Everyone makes errors, but here they own up to them in public! They then make their corrections easy to find. What's more, when they correct a web article, they clearly make a note of it with "Correction" in the title of the newly edited piece.

Likewise, Jackson says in his blog that there are multiple possible explanations for *why* this particular pattern evolved. It could be this, or it could be that. Generally speaking (and especially about evolution), leaving the door open to alternate explanations is a good idea—and also makes me trust the scientist more.

Research Lessons

1. *When something odd, peculiar, or funny strikes you (and of course, especially when it's central to something that you're trying to understand), it's worth doing just one more query to see if the odd or surprising thing holds up to a little scrutiny.* I advocate making this part of your daily inquiry practice; check out one "fact" a day. You'll be surprised at what holds up, and what doesn't. Just one more search to check; it's not much to ask.

2. *Take note of how reliable, consistent, and credible your sources are.* Sites that use inflammatory language usually have a position they're advocating; you don't need to ignore them, but you DO need to understand what their slant is going to be. And when you find high-quality credible sites, remember them. You'll come back to them in the future, and knowing which sites are reliable will improve your ability to triangulate between information sources.

3. *It's a good sign when publications admit their errors.* Be skeptical of sources that never admit to a mistake. Even the most careful scientists make mistakes and have to do an embarrassing public recantation. That's why I tend to believe the *New York Times* more often than not, and why I tend to believe other publications that clearly mark their errors in updates.

4. *One research question leads to another.* Here are the changes in my research questions just during this one investigation.

    1. *Is this fly for real, or is this all some kind of strange hoax?*
    2. *Does this fly actually have an image of ants on its wings?*

3. *Who are the people I found on my search? Should I believe their images and posts? What's their background?*
4. *Are those images of insects on the wings of this fly?*
5. *Why do you believe whatever it is you believe about this?*
6. *How variable are these kinds of fly wing patterns?*

Frequently, as I do online research (or as I watch expert researchers doing their work), the research question changes quite a bit from the beginning until the end. In this chapter, we saw the research question change from the open-ended one, "Is this fly for real?" to a more precise, "How variable are these kinds of fly wing patterns?" At each step along the way as I learned more and more about fruit fly wing patterns, I was able to refine the original question to be a little more exacting. Along the way I asked subquestions ("Who are the people I found on my search?" to determine if I could trust their writings) and even more general questions (I asked about whether they were "ants on wings" in the second question, but generalized it to "insects on wings" in the fourth). But each change in the research question was driven by what I'd discovered as a result of the previous searches. As you learn, you change your level of understanding as well as your questions. (Which is why I jot down what my research question is, just so I can keep track!)

5. *The path of research is often twisty. Don't panic; it's always that way.* While this chapter doesn't take long to reach a conclusion, this is actually the result of about four hours of my time (spread over a couple of days). I haven't shown you about all the dead ends—searches that didn't work, or websites I visited that were clearly bogus. This is just the way real research works: you explore, take notes, and back up when things don't work out. Sometimes you just have to be persistent, and that takes time. (And that's another reason why I write down my research questions. It sometimes takes me a day or two to get back to my online research, and seeing the note beside my keyboard reminds me what I'm trying to figure out!)

How to Do It

A. Search in this book. When you're doing a search in Google Books, you're usually looking for something that is in the text of the book.

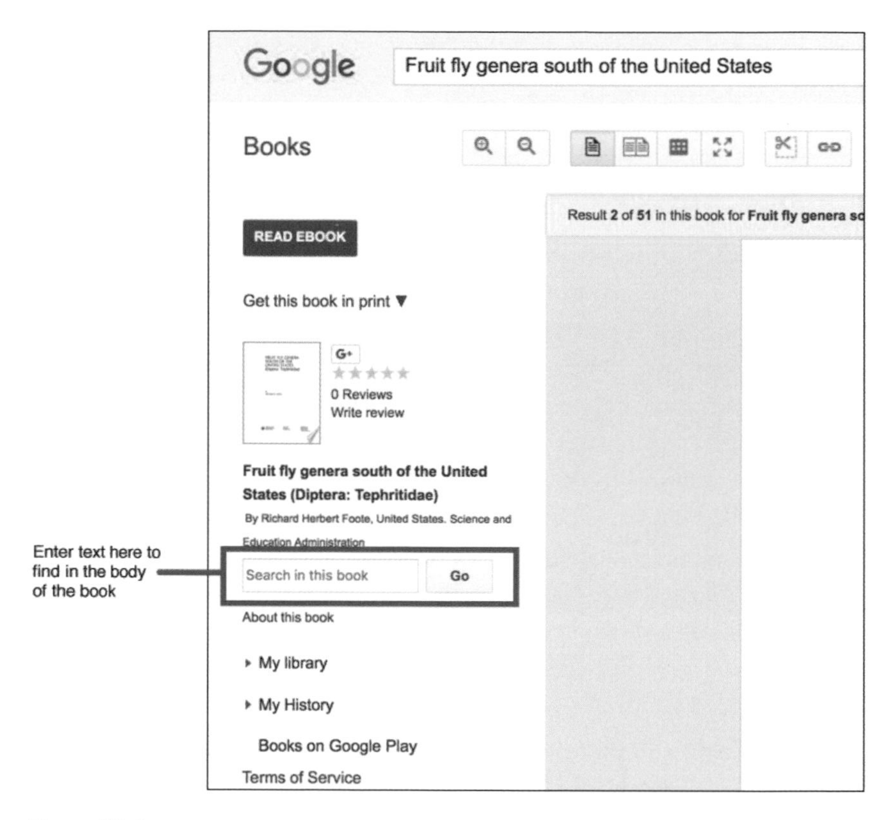

Figure 13.6
The "search in this book" box lets you find a bit of text in the body of the book.
Credit: Daniel M. Russell

For instance, if you use Books.Google.com and open up one of the books we used above (*Fruit Fly Genera South of the United States*), you should see the image in figure 13.6.

This search-in-this-book box is a useful tool when you want to locate a piece of text in the body of the book. In our case, we wanted to find different patterns on the wings of fruit flies, so I just entered the text *wings* into the search box, clicked on "Go," and found every mention of the word wings in the book. I scrolled down the list of hits and was able to see previews of the text surrounding the hit.

A few clicks later, and I had the illustration (and the information) that I needed.

Notice that a search for *wing* does not give the same results as for *wings*. Each word in the book is searched for separately. Unlike regular web search (which does synonym expansion), searching for text in Google Books is a search for exactly that word. (Which means that you should be sure your spelling is correct; it won't fix it for you automatically!)

Likewise, because of copyright restrictions, not all the text of all the books is viewable. Sometimes when you do a "search in this book," you'll find a hit that will show the term you're looking for, but not the page view. That part of the book is copyright protected and isn't viewable. If you really need to read that part, you'll have to buy the book or get it from a library. (Thank heavens for interlibrary loan!)

Try This Yourself

This chapter has been all about how to get to the bottom of a controversial topic, searching out and assessing different perspectives on it. It's often difficult to have these kinds of discussions on topics that are current events; frequently, those are surrounded by heated debates in which emotions run high and hot.

But this is an important skill to have, so we want to find a topic that has a diversity of opinions (all of which might be hotly debated, but people holding differing positions are still willing to sit together at a table and talk over their interpretations).

One such topic that is at a bit of a remove from current affairs is the interpretation of archaeological sites. There are plenty of sites that are clearly important, yet are difficult to interpret. Some are better known than others. Callanish is well known, but some of the lesser-known sites in Britain make equally fascinating research topics and have just as contentious interpretations, such as the Ring of Brodgar, Chysauster Ancient Village, or Avebury. Who built these remarkable sites? Why did they spend so much time and energy building them?

Figure 13.7
Callanish has been an enduring source of mysteries for centuries. It's one of a number of sites that are ripe for online research using maps, archives, books, and online resources.
Credit: Petr Brož, Wikimedia, CC-SA

14 What's the Connection between "The Star-Spangled Banner" and the General Who Burned the White House? How to Search for Vaguely Remembered Connections between Ideas

Finding the connections that you only vaguely remember. Ways to explore a serendipitous discovery, and when to stop.

It was a glorious spring day in Halifax, Nova Scotia, and the truth is that I was running, not whistling, through the graveyard on my afternoon run.

You might think that this is a tad disrespectful, jogging through the headstones, but I don't think those who have passed on mind much, and besides, this was an older cemetery, the kind that families in the nineteenth century would visit on the weekends for picnics and recreational outings.[1] We don't do that much today, so I figured that a great use for a cemetery is to run on through and say a quick hello to the headstones.

As I ran, I unexpectedly saw a grave marker that made me stop—literally—dead in my tracks, midway through my run. It was the final resting place of a man whose name I'd seen before, but who I did not know had ended up in central Halifax, under the leafy green trees of the Old Burying Ground.

For me, the best thing that happens when you're doing research is finding those unexpected connections that link things together. It's a magical moment when two different things suddenly turn out to be connected. The thing that stopped me while running was seeing that kind of connection in the engraved name on the stone.

I had discovered the final resting place of someone I'd read about before, via a name I recognized. Suddenly, two wildly different threads of my reading collided and merged. This man was involved in an interesting part of history, with another different yet equally interesting man.

But as I stood in the cemetery, dripping sweat, all I could remember was that these two were sworn enemies, and there was some connection, which I could not remember.

Just a couple of weeks earlier, I had been on another run through Georgetown, that leafy, tony area of Washington, DC, that has a wonderful jogging path that goes alongside a canal paralleling the Potomac River. Just before I reached the Key Bridge, I jogged through a small park with a bust of Francis Scott Key (figure 14.1). I snapped a quick picture with my phone's camera and kept running. (This happens a fair bit to me, and maybe to you: I use the camera on my phone to snap a quick picture as a way of reminding me of things I've seen and things I want to look up later.)

That evening I looked up a bit of the history of the park, learning that it is there, next to the (Francis Scott) Key Bridge over the Potomac, because this is where Key's house was from about 1805 to 1830, and also discovering that this is where he lived when he wrote the poem that became the US national anthem.

As I read about the Key's life, I was a bit surprised to learn that Key was actually stuck on a ship in the Baltimore harbor, watching, while the battle raged that he described in "O Say Can You See."

He had gone out into the harbor with fellow lawyer John Stuart Skinner (and also the US–British prisoner exchange officer). They would visit the British commander to negotiate the release of William Beanes, a local doctor who had been taken as a prisoner of war a few days earlier. It turns out that Beanes had been detaining and jailing British troops that were looting farms in his area. He didn't care for their looting, and not surprisingly, the British didn't care for his jailing of their soldiers. The British military were adamant, and since they had more guns when they stopped by his home late at night, the military tossed him in the brig, holding him prisoner.

During that meeting aboard the British ship, the HMS *Tonnant*, I discovered that Key and Skinner were the guests of three British officers: Vice Admiral Alexander Cochrane, Rear Admiral George Cockburn, and Major General Robert Ross—two naval officers and an army officer.

They negotiated over dinner for Beanes release, which they achieved. It was a refined negotiation, with wine.

After dinner, Key, Skinner, and Beanes, alas—sorry to say—were not allowed to return to land because they had learned the strength and position of the British units along with the British intent to attack Baltimore.

Figure 14.1
The bust of Francis Scott Key in the park located near where he lived in 1814, the year he penned "Defence of Fort M'Henry."
Credit: Daniel M. Russell

Thus, Key and Beanes were detained, and unable to do anything but watch the bombarding of the American forces at Fort McHenry during the Battle of Baltimore on the night of September 13, 1814. That must have been incredibly frustrating, but the experience led to the US national anthem as Key wondered at the survival of Fort McHenry and its flag, still flying in the dawn's early light. The resistance of Baltimore's Fort McHenry during the bombardment by the Royal Navy inspired Key to compose the poem "Defence of Fort M'Henry," which later became the lyrics for "The Star-Spangled Banner."

Yet it was that name—Robert Ross—that stopped me in Halifax. I recalled the name, but not exactly the connection (this was several weeks later, remember). Once again, I snapped a picture, and later was able to do a bit of research to reverse engineer my memory and figure out *why* finding Ross buried in Halifax was such a surprise. How did I know that name?

This moment of incomplete recognition in the Old Burying Ground led me to the following thought.

Research Question: *Who was Robert Ross? Why do I half remember his name?*

With the images I took, I could zoom in on the gravestone and make this transcription of the inscription on it (figure 14.2):

> Here on the 20 of September, 1814, was committed to the Earth the body of Major General Robert Ross who, after having distinguished himself in all ranks as an Officer in Egypt, Italy, Portugal, Spain, France, & America was killed at the commencement of an action which terminated in the defeat and rout of the Troops of the United States near Baltimore. On the 12th of September, 1814. At Rosstrevor, the seat of his Family in Ireland a monument more worthy of his Memory has been erected by Noblemen and Gentlemen and the officers of a Grateful Army which under his command attacked and dispersed the enemy at BLADENSBURG on the 26th of August 1814 AND on the same day victoriously entered Washington, the capital of the United States. In St. Paul's Cathedral a monument has also been erected to his memory by his country.

That's all interesting information, but why do I have a half recollection of his name roaming around in my brain?

One way to bring out a partial memory is to quickly scan a bunch of articles about the subject and see what strikes you as memorable. With luck, you'll find something that tickles your memory and be able to recall the connection.

In this case, I did a quick search for [Major General Robert Ross] and brought up many articles on him, including an interesting observation that he is best known for the burning of Washington, which included the burning of the White House, Capitol, and Library of Congress on August 24, 1814. That's certainly a reason why I might have particularly remembered his name. Not only did his name surface in connection with Key aboard the HMS *Tonnant* but he was also the major who led a successfully infamous attack on Washington, DC.

Figure 14.2
A close-up of the grave marker that I found in Halifax
Credit: Daniel M. Russell

If you read the gravestone inscription carefully, you'll see that he was buried here in Nova Scotia on September 20, 1814, less than one month after his forces destroyed the White House. Oddly, he was actually killed in action eight days earlier on September 12, 1814. (And the story goes, he was transported from near Baltimore, where he was killed, to Halifax in a barrel of rum.[2] It was a busy summer for him.)

Scanning the search results, I also read that he was killed during the Battle of Baltimore at North Point, *not far from Fort McHenry.*

Reading that little bit of text made the connection for me: in my mind, Ross is associated with the burning of the White House, Battle of Baltimore, and assault on Fort McHenry, which led to "The Star-Spangled Banner" and hence Key. When I saw the mention of Fort McHenry, the name of Ross and the fort clicked together, and I remembered that this was the link.

Q: How can I find more about the connection between Ross and Key?

To search for this, I did the kind of search query you might do to find a connection between any two people, putting the two names side by side in a single query. Here I've put both names in double quotes to search for each name separately (I don't want to find out anything about someone named "Key Robert"):

["Francis Scott Key" "Robert Ross"]

This leads to a bunch of content, including several books about the War of 1812. Here's one descriptive passage:

> It was aboard the Tonnant that the Americans Colonel John Stuart Skinner and Francis Scott Key dined with Vice Admiral Cochrane, Rear Admiral Sir George Cockburn, and Major General Robert Ross, where they negotiated the release of a prisoner, Dr. William Beanes. After his release, Skinner, Key, and Beanes were allowed to return to their own sloop but were not allowed to return to Baltimore because they had become familiar with the strength and position of the British units and knew of the British intent to attack Baltimore. As a result, Key witnessed the bombarding of Fort McHenry and was inspired to write a poem called "The Defense of Fort McHenry," later named "The Star-Bangled Banner."[3]

If you read the preceding quote, you'd (naturally) assume that these gentlemen had dinner, and then Skinner, Key, and Beanes went back to their sloop to watch the bombardment of Fort McHenry.

But that would be wrong!

I was a bit curious about the Battle of Fort McHenry, so I did a search to get a bit more background:

[Battle of Fort McHenry]

In these results, I find that the battle (that is, the all-day-all-night bombardment of the fort from the British ships) happened on September 13–14 starting at 6:00 a.m.

But wait a second! Wasn't Ross killed on September 12? (That's what his gravestone says, and that's what all his biographies say!) How is it possible that he died on September 12? Wasn't that the day that Key and Ross had dinner together?

To answer this question, I needed more details, which led me to search in Google Books using that same query ["Francis Scott Key" "Robert Ross"] that we used before, looking for books that would tell the story in a bit more detail. That is, I was hoping for a book that would give a detailed chronology of the last few days of Ross's life. This time I checked a few more books before I found what I needed.

In one of those books, *What So Proudly We Hailed: Francis Scott Key, a Life* by Marc Leepson, I **searched inside the book text**[A] and found that this fabled dinner took place on September 7, at the mouth of the Potomac River, which isn't exactly near Fort McHenry. That's a bit of a surprise as it's neither the date nor place I expected based on what I'd read so far.

In Leepson's book, we find that

> the three Americans were first escorted to the British frigate *Surprize*, under the command of Admiral Cochrane's son, Sir Thomas Cochrane. On September 10, Key, Skinner, and Beanes were allowed to return to the *President* [their boat], accompanied by British marine guards. The *Surprize* [the British boat] towed them into Baltimore Harbor, about eight miles from Fort McHenry, behind the fifty-ship British fleet, where they dropped anchor. They spent the next four days there.[4]

Basically, they were waiting for the attack on Baltimore to finish up so they could go home. The British really wanted to take Baltimore, which they referred to as a "nest of pirates" because of the strategic location of the Baltimore harbor, and the number of Baltimore clippers (a kind of fast ship that the Americans used) along with their tendency to raid, seize, and sink British ships while on the high seas.

In the meantime, in the early morning of September 12, Ross led his expeditionary force onshore at North Point, trying to run around the flank of the Americans that were dug in all around the city.

There, about seven miles south of the city, he was killed by an American bullet just after he rode to the front to monitor the skirmish.

As this map shows, while all this was going on, Key was anchored a few miles offshore with a great view of Fort McHenry and its bombardment by Congreve rockets ("*rockets red glare*") from the rocket ship HMS *Erebus*.

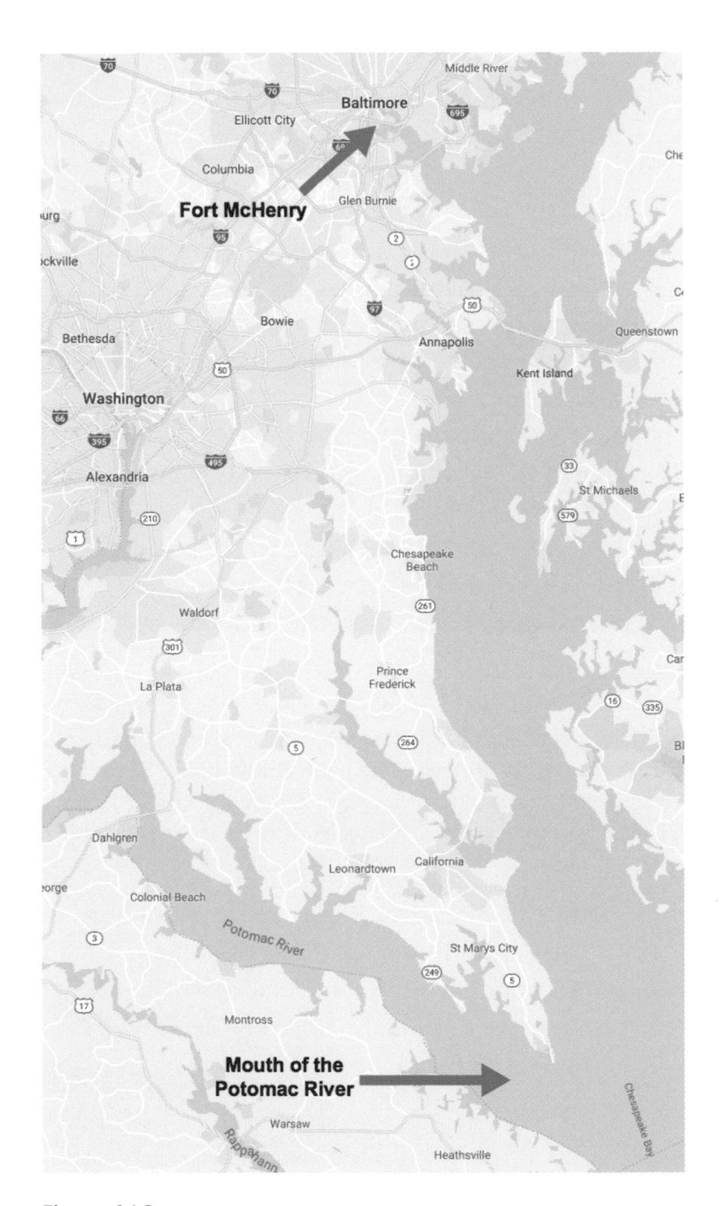

Figure 14.3

A map showing the relative location of the place where the parley meeting took place (at the mouth of the Potomac) and where Key watched the Battle of Fort McHenry. Credit: Map data © Google; edited by Daniel M. Russell

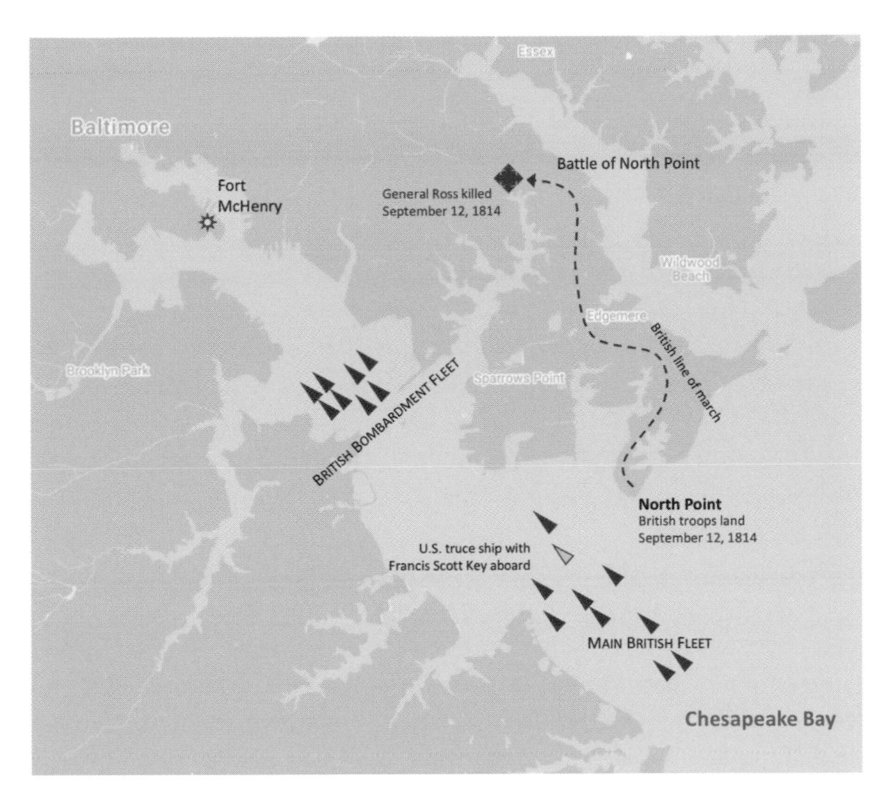

Figure 14.4
Map of Fort McHenry and the location of the British fleet. Note that Ross was killed well to the east of the fort on September 12, 1814.
Credit: Daniel M. Russell

Bomb ships HMS *Devastation*, *Terror*, *Volcano*, *Aetna*, and *Meteor* ("*the bombs bursting in mid-air*") all shot their mortar shells toward the fort.

So while Key was stuck on the *President* in September 1814, he wrote out the poem "Defence of Fort M'Henry" with the melody of a popular drinking tune in mind ("To Anacreon in Heaven").[5] Over the next few years, it quickly became a popular patriotic song to sing, much as "God Bless America" is a popular patriotic song in our day. It wasn't until 1931 that the US Congress made "The Star-Spangled Banner" the national anthem of the United States.

How to Find Serendipity While Avoiding the Rathole

Serendipity strikes when, in the process of looking for one thing, you find something else that's a happy accident. It's a bit like looking for a flashlight in the kitchen junk drawer and coming across a hundred dollar bill: you weren't looking for it, but it's a happy moment when you discover it.

Likewise, serendipity happens when you're doing online research and stumble across something you really didn't expect to find. But as Louis Pasteur once wrote, "Chance favors the prepared mind."[6] And in the case of research, "prepared" means that you actually know something about a topic—that is, you recognize a little something in the texts you're reading that sticks out, even if only partly. In order for that to happen, you *already have to know something* about the topic. Serendipity doesn't strike the mind that doesn't know.

That was the case for the original question in this chapter; I'd been reading about Key but when I ran past that gravestone in Halifax, I partly recognized the name, which led me to this entire line of research.

So with a prepared mind, as I was finishing up my search for Ross and Key, I couldn't help but note that Key had an extraordinary life. He was a fascinating man—a lawyer (who assisted in the conspiracy trial of Aaron Burr), poet, founder of two seminaries, and antiabolitionist who nevertheless freed his own slaves.[7] In reading about his life, I ran across something else that piqued my interest. Strangely, his son, Philip Barton Key II, was shot and killed by a US congressman from New York in 1859. Earlier this year I'd been reading about the use of the insanity defense in murder cases, and once again, something in this name made my inner ears perk up with interest.

Was there really a connection here? Amazing. How could I not find out more about this serendipitous discovery?

Rather than read about this salacious story as a footnote to the life of Francis Scott Key, I searched for his name directly: [Philip Barton Key] landing on his Wikipedia page.[8] There I noticed a link at the top of the article: *"For other people with the same name, see Philip Key (disambiguation)."*

When you see this, *pay attention*, because it's probably worth checking it out, as it means that there is more than one person with this name.

Philip Barton Key II

From Wikipedia, the free encyclopedia

For other people with the same name, see Philip Key (disambiguation).

Harper's Weekly engraving of Philip Barton Key from a photograph by Mathew Brady

Philip Barton Key (April 5, 1818 – February 27, 1859)[1] was an American lawyer who served as U.S. Attorney for the District of Columbia.[2] He is most famous for his public affair with Teresa Bagioli Sickles, and his eventual murder at the hands of her husband, Congressman Daniel Sickles of New York. Sickles defended himself by adopting a defense of temporary insanity, the first time the defense had been used in the United States.[2][3]

Figure 14.5
When you see a link on Wikipedia that says "(disambiguation)," you probably want to check it out. This means that there are multiple people with the same name who can be easily confused with each other.
Credit: Wikipedia

Clicking on this link told me that Key II was indeed the son of Francis Scott, and the great nephew of Philip Key, the US congressman from Maryland (1791–1792).

But I also found several other articles about Key II, all of which agreed that he was in fact murdered by Senator Daniel Sickles.

It turns out that Key II was having a rather-public affair with Sickles's wife, Teresa.[9] In a fit of jealous rage, Sickles shot Key II dead just outside his home, in Lafayette Park, right across Pennsylvania Avenue from the White House! (And you thought today's homicides were ridiculous.)

As I read more about Sickles's life and his murder trial, I learned that he was acquitted because he was judged to be momentarily insane. This was the vague recollection in my memory that once again sent me down a research path.

A quick search for [Daniel Sickles reason of insanity] to confirm this shows that he *was* the first person to use the "murder by reason of insanity" defense in the United States. Remarkably enough, he was arrested and put

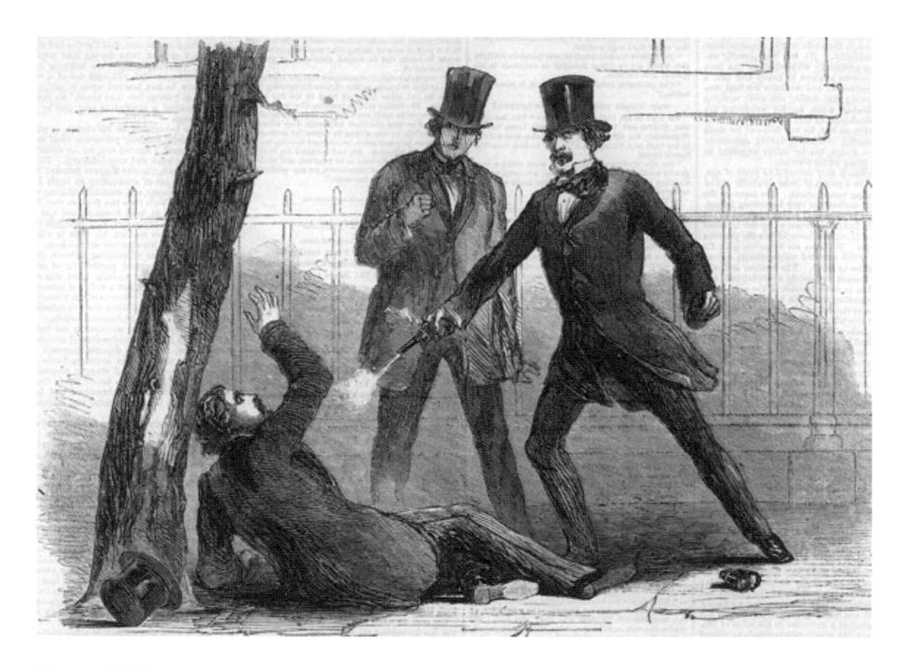

Figure 14.6
Senator Daniel Sickles murdering Philip Barton Key II in downtown Washington, DC.
Credit: Library of Congress Prints and Photographs Division, Washington, DC; original Illustration in *Harper's Weekly*, March 12, 1859, 169.

into jail, but was allowed to receive visitors AND keep his sidearm—while in jail. (I guess security was a little more lax in those days.)

This search tells us all kind of things—about Sickles's tendency to squire and gad about with a well-known prostitute on his arm.

Out of the wealth of content available online, the richness, complexity, and pure salaciousness of his story, learning about Sickles is probably best done by reading a fully researched book on his life. Perhaps the best book on Sickles's amazing, checkered, and complicated life is Thomas Keneally's *American Scoundrel: The Life of the Notorious Civil War General Dan Sickles.*[10]

After this rather-tawdry affair with the murder of Key II, trial, and acquittal by reason of insanity, Sickles went on to become a general in the Civil War. He was a general who fought, rather badly it seems, at Gettysburg, where he managed to get almost all his brigade killed by not following orders.

You could continue in this vein, learning about Sickles's wife (who apparently spoke four languages), mistress (Fanny White, who he presented to Queen Victoria), or time in Spain (when he was rumored to be the paramour of the deposed queen). The stories are fascinating—and endless.

As it is with many things on the internet, you have to know when to quit. Although the Sickles, Key, and Ross interconnections and stories are intriguing, an important skill when doing online research is being able to stop. If you're having fun and have the time, by all means, keep reading! But more often than not, there's a time limit on how long you can spend following the trail of scandal and strangely compelling life stories.

And that's why I write down my research questions. Even just jotting them down on a piece of paper gives you a bit of guidance. Once you've found the answers to those questions (that you DID write down, yes?), consider stopping your searches, lest you keep wandering around in the deeper recesses of the internets. As a positive side effect, when you've got the original research question written down in front of you, you tend to stay on topic. (Although staying on path sometimes comes with a little ping of recognition—oh, *that's* what I'm searching for!)

Another trick I sometimes use is to set a timer using my online calendar and then pay attention when the time I allocated for doing the research is over.[11] The trick here is to *really stop* when the timer goes off (or the calendar event pops up). I've found through hard experience that it's too easy to just "do one more search" or read "just one more article."

Research Lessons

1. *I've said this before, but it constantly bears repeating: READ CAREFULLY!* I have to admit that I misread those passages about the dinner between Key and the British when I was first looking into these historic folks. I initially jumped to the conclusion that the dinner was on the same night as the famous attack. It was only later (as I was writing out my notes) that I noticed that Ross actually died the day *before* the bombardment. That contradiction made me back up and reread the original material, looking for the details.

2. *The more you know about a topic, the more likely it is that serendipity will help you out!* Serendipity happens to those who know. If you've already read

fairly widely on a topic, the likelihood that you'll see something that's useful to your research goes way up. In a sense, this is the rich-get-richer effect (aka the Matthew effect), except in terms of serendipity, it's a case of "the more you know, the more you can learn."[12] A good heuristic to enhance serendipity is to read widely in the topics in which you want serendipity to happen.

3. *Write down your questions to keep on focus.* I often just scribble them in a notebook that I keep beside my keyboard. That way I'm always reminded of what I'm searching for; it keeps me oriented and on task. It's simple, it's dumb, and it completely works to keep me on point.

4. *A useful trick to get to a deeper understanding is to write up a small essay that presents all your information.* I'm 90 percent convinced that having to write something down (and have that write-up make sense) is a great method to making sure that all your ducks are in a row. If you're being honest with yourself, you'll pick up all KINDS of mistakes in your reasoning and data (such as when it seemed that Ross was killed before the date that I'd assumed was the day of their dinner!). The simplest, fastest way I know to write a short essay is to send an email to a friend who shares your interest. In my case, I bombard friends with emails that start out "I bet you didn't know that …," and then I write on and on about, for instance, finding Ross's grave in Halifax and telling the rest of the story. The key thing is that in order to write that short essay (or email), you have to make sense of what you've found. You won't just write gibberish to a friend (at least, I don't recommend that), but you'll see any gaps or inconsistencies in what you've discovered.

5. *STOP when you run out of time or find the answer to your questions.* Research can be like potato chips—you want to keep going. But keep your head up and pay attention to how you're using your time. That's why I often use a timer when I'm in a hurry; it prevents me from overrunning the amount of time that I have and makes me refocus on the task.

How to Do It

A. Search inside the book text. Once you've found a book, you can search within the book for exactly the text you seek. In this case, I found the book *What So Proudly We Hailed: Francis Scott Key, a Life*, and wanted to find out

Figure 14.7
When you've opened a book, you can search through its contents by entering your search terms in the box that has the text "Search in this book," which lets you find specific words that are used in the text.
Credit: Google and the Google logo are registered trademarks of Google Inc., used with permission

what the author had written about that fateful dinner where Key, Beanes, and Skinner had dined. To do this, I found the book and then used the "Search in this book" search tool, handily located in the middle of the right-hand side of the book interface, just below the title.

In this case, I did a "Search in this book" for [Surprize] since I had already read that this was the ship where they'd had dinner. I could have also searched for their names, but I thought that their names would appear everywhere throughout the book, while the ship's name (with an unusual spelling) would appear only once in the text, just when and where they dined.

Try This Yourself

Everyone and everything seems to be connected these days. This is so true that the Wiki Game has sprung up to take playful advantage of this fact.[13]

But in online research tasks, the goal is usually to get to an understanding of the research question as rapidly as possible. The ongoing question for you, the researcher, is this: What should I do next? That is, should you modify the query, open up a new line of inquiry, try looking at some images, or just stop the research altogether.

I find some research questions so fascinating that I spend more time than I should reading *around* the topic. While that frequently provides useful context, if you're on a deadline, you need to be efficient in your searching.

So when do you stop? Generally speaking, when you stop learning useful new information about your question. Practically speaking, that can be tricky.

Here are a few research questions that captured my imagination for longer than they should have. As you work on these, continuously think to yourself each time you do another query, *Will this help get me to the goal? Will it help me answer my research question?*

1. *Heroes*: Who was Tadeusz Kościuszko, and how is it possible that he became a national hero in four different countries? He must have an interesting life story to become so celebrated in so many places.
2. *Studying climate change through artistic models*: A large collection of undersea invertebrate animals created in the 1860s is being used to study the effects of climate change. Who made those models? Are they really accurate enough to be used for this kind of comparison study?
3. *Desserts around the world*: Open-ended questions are often the most difficult to manage, mostly because the question isn't well defined. When you're doing research on an interesting topic, watch out! You might well follow those fascinating threads of research for hours.

Recently I got lost (for several happy hours) in the topic of the most popular desserts in the world. I quickly realized that each country has its own list, but that there are lots of overlaps that I hadn't expected. How, for instance, does a plant native to Brazil turn into a common dessert in India, Thailand, Congo, and the United States? Tapioca pudding is often considered a somewhat-bland dessert choice in the US Midwest. How did it get to be thought of as a modest dessert in that region?

As an alternative question to search, finding the connection between different desserts can be seductive. For instance, what's the connection between the French mille-feuille and Middle Eastern baklava? (And who invented baklava anyway?)

15 What Causes the Barren Zones around Some Plants? How to Know When You Should Go Offline and Do Research in the Real World

Some of the brush in chaparral lands have an odd zone around them that lacks small plants. That's strange. Why? A story about finding the answer from a number of different sources, some of which are scholarly, and why you should use them in your research too.

The biggest surprises are sometimes the ones that you don't see at first. They're the little things that slowly creep into your consciousness and make you wonder, Why is this so? It's only after some time that you realize how big of a surprise it was. This is one of those retrospective surprises.

Not long ago on a beautiful summer morning, I went for a run up on Black Mountain, one of the local peaks in the Santa Cruz mountain range. It hovers over Silicon Valley, making up most of the western edge of the mountain range that turns Silicon Valley into a real valley. If you look carefully on the right side in figure 15.1, you can see the typical summertime wave of fog coming in from the Pacific Ocean. On typical summer evenings, that fog (aka "Karl," as the locals affectionately call it) comes in from the sea to cover San Francisco. Luckily, Black Mountain is usually high enough that the fog comes around, not over, the peak.

When I was there, I noticed something that was a tiny yet odd effect. The mountain is part of an open-space preserve, so there are the usual trees, trails, wandering deer, and chaparral. In the chaparral, there is a lot of brush. Here are two pictures I took on that run of a strange phenomenon (figure 15.2).

What struck me about each of these pictures is that there's a noticeable gap between the green plant and the grassland next to it. Oddly, this doesn't happen ALL the time, but it does happen a lot.

Figure 15.1
Black Mountain
Credit: Daniel M. Russell

Figure 15.2
Credit: Daniel M. Russell

What's going on here?

I know the plants here are chamise (*Adenostoma fasciculatum*) and coyote brush (*Baccharis pilularis*), living side by side in large brushy zones.

But looking at these plants now was odd. Noticing this strange, plant-free boundary around most (yet not all!) of the plants on Black Mountain led me to ask the following research question:

Research Question: *What causes the odd, plant-free gap around the plants in these photos?*

Is there some way to figure this out with online research?

If it helps (and it might or might not), these photos were taken on August 13, 2013, at latitude/longitude 37.3195203, –122.1630756. Here's what the place looks like in a Google Earth image (figure 15.3).

Look carefully at this brushy area marked by the red drop pin. I know that this is made up of chamise and coyote brush. When I was visiting, I

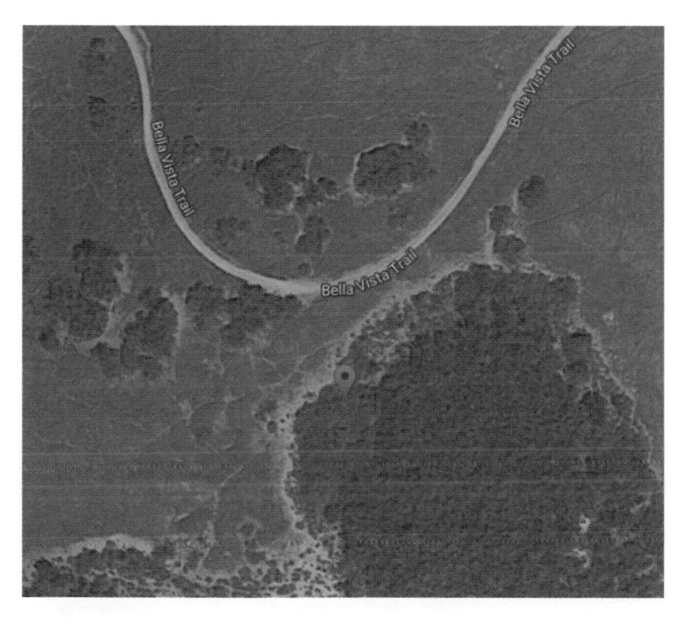

Figure 15.3
Typical brush complexes on Black Mountain. Even from this altitude, you can see open ground around most of the clusters of brush. I wonder why this is so?
Image © Google

noticed that you can see this kind of gap all around these plants. (The light-colored perimeter is where you can see the soil.)

What's going on here? I looked carefully, and this gap at the periphery isn't a trail of people or deer. (You can see deer trails weaving between the clusters of bush, especially to the west of the drop pin.)

So what's causing this?

I began my search with a Google query like this:

[plant free zone OR area near chamise OR "coyote brush"]

Yet these results weren't especially good. Looking at the search results, I learned a lot about how to create "plant-free zones" around houses, especially those that are built in chaparral or near fire-prone plants like chamise. From the US Forest Service website, I also learned a great deal about chamise, such as that

> under natural conditions, dormant seeds accumulate in the soil until stimulated by fire to germinate. Chamise seeds are unpalatable and seedbanks apparently are not subject to heavy predation. Consequently, chamise seed densities increase over time. Seed density in the seedbank beneath 9-year-old stands was estimated at 2,000 seeds per square meter while in 85-year-old stands, seed density was approximately 21,000 seeds per square meter.[1]

Reading through these articles was interesting, but not especially useful. (Or so I thought. See below.) I needed to try a different research tack.

I completely modified my query to focus in on just chamise:

[plants not grow near chamise]

Now I had a bunch of new, additional results just about chamise. But one result struck my eye: a PDF from the Santa Cruz County Parks Department that said (in the snippet), "Chamise uses allelopathy, altering the soil as it drops leaves." This sounded as though it MIGHT be relevant, so I clicked through to the landing page and found these lines that sound like our solution: "Some plants engage in chemical warfare, or allelopathy. Chamise accumulates toxic, water-soluble compounds as a result of normal metabolic processes. Fog drip and rain carry these toxins to the soil, where they inhibit the growth of competing plants."[2]

Now this sounds like the right mechanism; this well might cause a zone around the chamise that would be pretty barren. This paragraph, however, uses a word that's new to me: "allelopathy." I already knew that *something* +

pathy usually means that something was being killed off. Could it be that the seeds or plants near the brush was being killed? That was enough of a hint.

I looked it up. Even though I'm a professional researcher, I don't know the meaning of every word, but this seemed promising. "Allelopathy" means "the chemical inhibition of one plant (or other organism) by another, due to the release into the environment of substances acting as germination or growth inhibitors." This well could be exactly what's happening to cause the plant-free zone around the bushes. Broccoli and black walnut, for instance, both exude substances that kill (or prevent from growing) plants living nearby. So maybe we've found the answer.

And now that I know this word, I remembered seeing it before—in the first set of results! Now I know what that word means and see in retrospect that my first hint was actually in the first search I did, if only I'd known.

This is now officially interesting, so I dug a little deeper into the allelopathy vein.

I thought what I'd do is to search for "allelopathy" for chamise, with the obvious query:

[allelopathy chamise]

Not only did I find a lot of new results, but I found myself in the middle of a scientific battle of explanations!

Here is the results page (figure 15.4).

As I read the first result, I was astonished: this link gives you an idea about what's going on. In that article, published in 2004 (as part of the oldest botanical publication in the United States, the *Journal of the Torrey Botanical Society*), the author *calls into question the idea that chamise is allelopathic.*[3] So I did a bit of searching and reading of these pages. I found out that on the "allelopathy side," there are several publications arguing that chamise IS allelopathic and causes the plant free zone around chamise.

Keeley, J. E., B. A. Morton, A. Pedrosa, and P. Trotter. 1985. "Role of Allelopathy, Heat, and Charred Wood in the Germination of Chaparral Herbs and Suffrutescents." *Journal of Ecology*, 445–458.

McPherson, J. K., and C. H. Muller. 1969. "Allelopathic Effects of Adenostoma Fasciculatum, 'Chamise,' in the California Chaparral." *Ecological Monographs* 39 (2): 177–198.

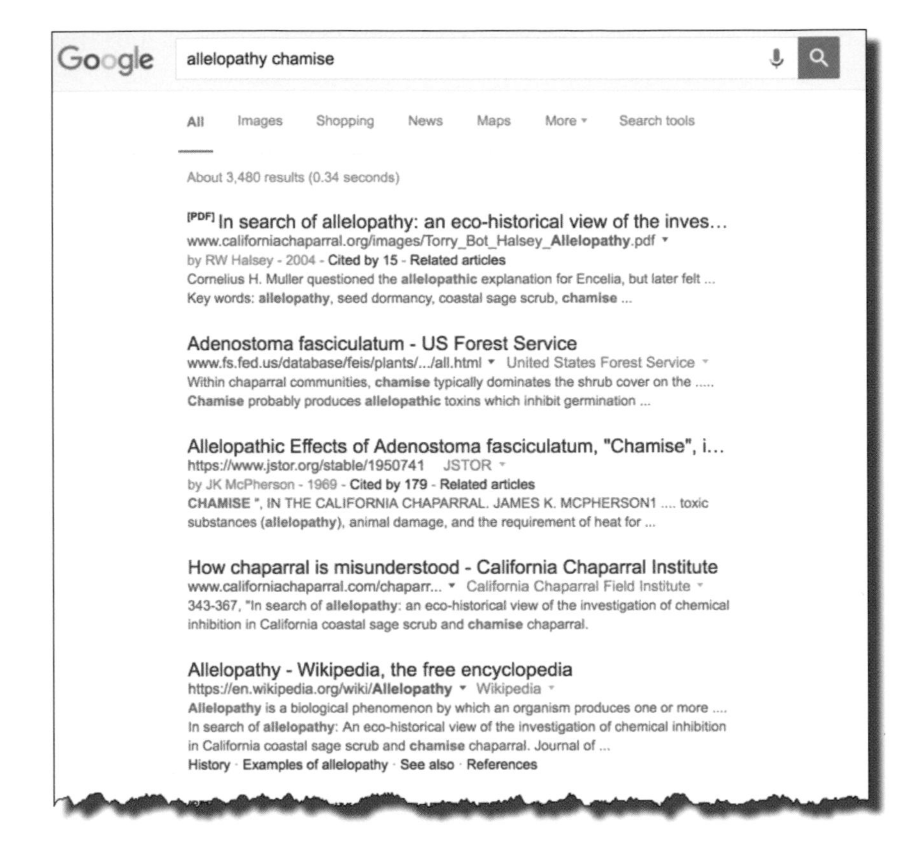

Figure 15.4
Credit: Google and the Google logo are registered trademarks of Google Inc., used with permission

Muller, C. H. 1953. "The Association of Desert Annuals with Shrubs." *American Journal of Botany* 40:53–60.

Muller, C. H., R. B. Hanawalt, and J. K. McPherson. 1968. "Allelopathic Control of Herb Growth in the Fire Cycle of California Chaparral." *Bulletin of the Torrey Botanical Club* 95, no. 3 (May–June): 225–231.

Notice that these publications are all a few years old. Science marches on, discovering new things all the time. Is the information in these articles still widely believed to be true? We should look for more recent work.

On the other hand, there are a few more recent publications that argue this is all wrong, that allelopathy isn't really what's going on here (at least for chamise):

Halsey, R. W. 2004. "In Search of Allelopathy: An Eco-Historical View of the Investigation of Chemical Inhibition in California Coastal Sage Scrub and Chamise Chaparral." *Journal of the Torrey Botanical Society*, 343–367.

Weidenhamer, J. D. 1996. "Distinguishing Resource Competition and Chemical Interference: Overcoming the Methodological Impasse." *Agronomy Journal* 88 (6): 866–875.

After reading a bunch of these papers, it becomes clear that the idea that chamise is allelopathic originated in Cornelius Muller's original paper from 1953. Since then, though, the idea that chamise actively emits chemicals (or more likely, that the chemicals are leached off in rainwater over the plant) has been called into question.

There's no doubt that some plants do emit compounds that suppress nearby growth, but the question being debated here is whether or not chamise is one of them. The articles are fairly detailed, with long sections describing how to test one idea or the other. It's fascinating reading, if you're into details of methodology (which I actually quite enjoyed, although maybe that's just me), but I'll summarize it by saying that these folks are still trying to figure it out.

Nevertheless, a nice summary of this entire debate is contained in a relatively recent book, *Perspectives on Plant Competition*, published in 2012. It concludes that allelopathy by chamise and other chaparral plants is "overrated ... if not altogether fictitious."[4]

Why this might matter is because some botanists are trying to figure out ways to harness plants' natural allelopathic capability to do weed suppression in fields and perhaps farm without using any weed killer sprays on the crops. In essence, if chamise IS allelopathic, might it be possible to breed (or genetically engineer) a food plant that needs no weed killer, but can grow just fine on its own?

YET while reading *Perspectives on Plant Competition*, I looked up each of the citations in Muller's work and in particular each reference in the text about allelopathy. In the process, I found this remarkable phrase:

> [Here is some] experimental evidence supporting an alternative explanation for the bare zones around chaparral shrubs. Batholomew (1970) demonstrated that halos around some shrubs in the soft chaparral might be maintained by herbivory from animals inhabiting the shrubs. Successive attempts to test the relative importance of allelopathy versus herbivory in both the development and the maintenance of bare zones showed that chemical inhibition occurred in some

> species, but not in others and that the herbivory of unexclosed seedlings could be an intense as 100% (Halligan, 1973, 1975, 1976; Christiensen and Muller, 1975).[5]

I know that's dense prose, but WOW! If you decode this text, it's saying that animals are eating the small plants around the shrubs. That's what "herbivory" means: animals eating plants. "Unexclosed" sounds like a double negative, but it really just means that there's no animal-excluding fence around the small plants just outside the shrub. And not just some of the plants but instead up to 100 percent of them!

Looking up the Bartholmew reference, "Bare Zone between California Shrub and Grassland Communities: The Role of Animals," we find that

> previous studies have emphasized the role of volatile inhibitors of plant growth in producing this bare zone. However, there is a concentration of feeding activity by rodents, rabbits, and birds in this zone; if this activity is prevented by means of wire-mesh exclosures, annuals grow in the bare zone. Thus, animal activity is sufficient to produce the bare zone.[6]

Among other things, we now have a technical term for this region around the outside of chamise and coyote brush: bare zone. My next query was:

["bare zone" chaparral]

Voilà! We have a lot of articles on the bare zone around chaparral plants, just as we're seeing in the photos. The consensus of opinion now seems to be that the bare zone is caused by animals that are foraging near the clusters of chaparral shrubs. (This is supported by a bunch of those papers, specifically: Bartholomew 1970; Halligan 1974; Bradford 1976; Barbour et al. 2000.)

In this long list of articles, there's a wonderful (and readable) one from *Bay Nature* magazine titled "A Landscape Shaped by Fear on Mount Diablo." In it, the author writes,

> Muller's work was later challenged by researchers who argued that his work glossed over the role of hungry seed and plant eaters. "In California, the chaparral and coastal sage shrubs form excellent cover for rodents, rabbits, and birds," Stanford graduate student Bruce Bartholomew wrote in a seminal 1970 paper.
>
> These small animals leave the shelter of chaparral in search of food. But they don't go far, lest they themselves become a meal for hawks, bobcats, mountain lions, coyotes and other predators. That's why the bare zone is also called the "scurry zone."

> Where Bartholomew set up cages excluding small animals, the bare zone largely disappeared. That and followup experiments led to the view held by many field biologists today, that wary little animals play a huge role in keeping chaparral-grassland borders free of vegetation. In their view, the bare zone is not so much a strip of dirt poisoned by plant chemicals, it's a landscape shaped by fear.[7]

The article is well worth a read. In it, you meet University of California at Berkeley botanist Lindsey Hendricks-Franco and get a great picture of the exclosures that she's set up to test her hypothesis that small animal herbivory at the margins of brush zones in the chaparral causes bare zones there.

Of course, not being satisfied with what I'd read, I decided to go back out to my local chaparral and do a few of my own observations. Here are two photos that are pretty suggestive of what's going on (figures 15.5 and 15.6).

In figure 15.5, you see a bare zone next to a large stand of coyote bush. So far as I know, there hasn't been any report of allelopathy for coyote bush. If that's true, what's causing the bare zone next to it? Probably the small animals that live within.

Meanwhile, just down the trail I saw a bunch of grass growing right next to a small patch of chamise. If it was truly allelopathic, why would the grass be growing there (figure 15.6)?

Note that the stand of chamise is pretty small. I noticed that *small* patches of chamise OR coyote bush don't have bare zones, while *large* patches of chamise OR coyote brush do have them.

I speculate that a smallish stand-alone chamise bush just can't provide the cover that large patches can. That's why small bushes don't have a bare zone; if you're a mouse, there's no protection. But a big stand of any kind of chaparral brush will do: chamise, coyote bush, or even manzanita (another common California chaparral bush).

The Google Earth photo (figure 15.7, from March 2013) shows this gap effect quite clearly. Each of these clumps of bushes (some are chamise, and some are coyote brush) has a small zone around it that clearly shows either the allelopathic effects of chamise or the herb-grazing effects of small mammals near the bushes. (Note also the change in the color of the soil near the bottom of the image; it really does change to a light color down there. I checked it out in person. This isn't an artifact; it's really a color shift of the soil.)

Figure 15.5
A bare zone beside coyote bush (which nobody believes has any allelopathic effects on nearby plants).
Credit: Daniel M. Russell

And if you look at many of the clumps of bushes, you can see a pretty clear effect of the presence (or absence) of a bare zone depending on the size of the clump of bushes (figure 15.8).

Once the clump gets below a certain size, the bare zone seems to disappear.

After all this online research, I decided to go check out the truth on the ground. A few weeks later I went back up to the place where I took those pictures above. This place is Black Mountain (latitude/longitude 37.319795,

Figure 15.6
Grass growing immediately next to a small patch of chamise. Why isn't there a plant-free zone around this?
Credit: Daniel M. Russell

–122.163153), a mixed oak woodland, chaparral, and grassland in the Santa Cruz mountains, just west of Silicon Valley in California.

In figure 15.9, we have an overview of the same region we saw before, just zoomed out a bit to show more of the land around the clumps of brush.

I had speculated that small bushes couldn't provide enough cover for small critters to live in (and therefore wouldn't have a barren zone around them).

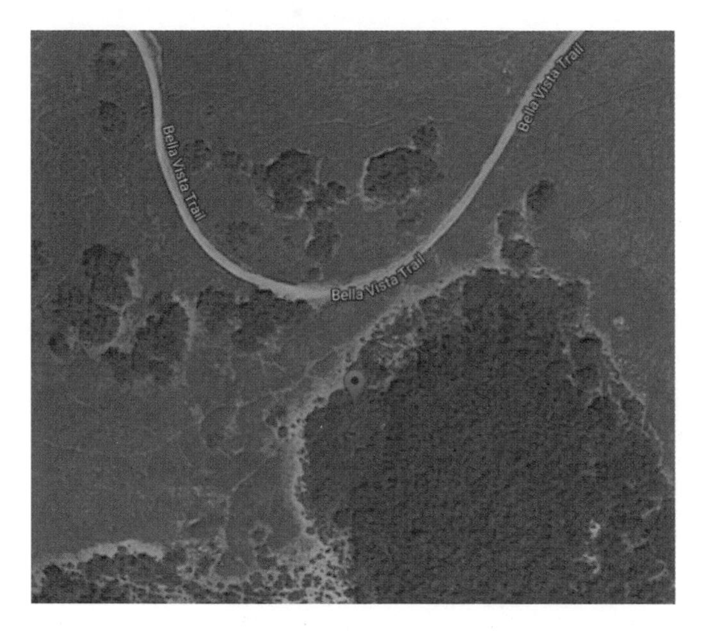

Figure 15.7
Image © Google

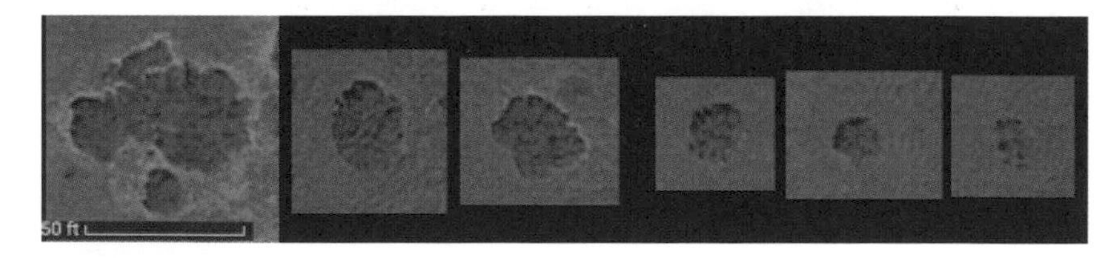

Figure 15.8
Bare zones seem to vary by bush clump size (and not by what kind of bush it is). These clumps are all taken from the area just to the north of the satellite photo shown above. The three on the right, being fairly small, show no bare zone.
Image © Google; edited by Daniel M. Russell

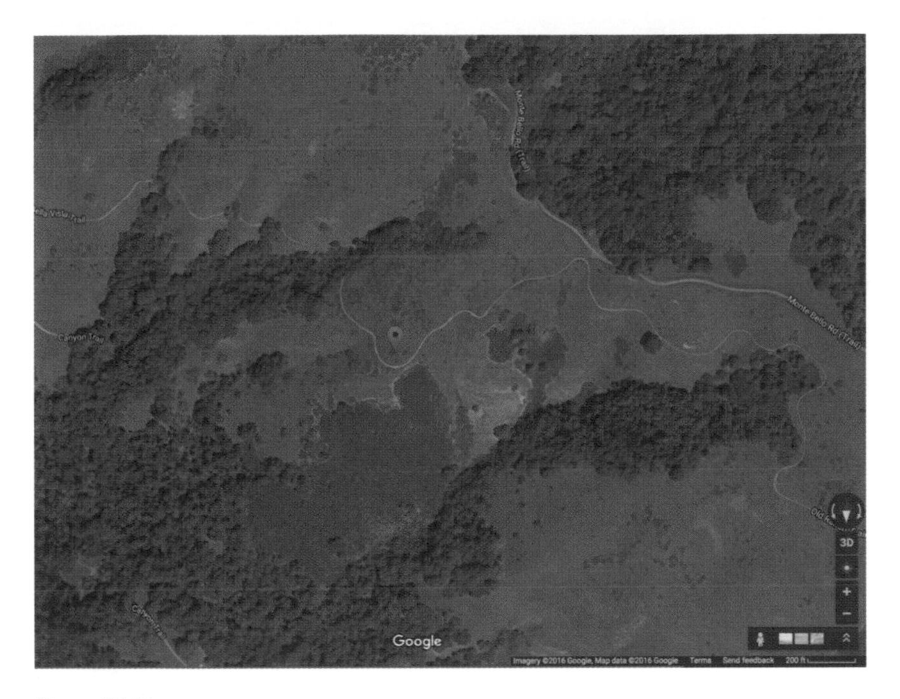

Figure 15.9
Overview of barren zone site on Black Mountain.
Credit: Map data © 2016 Google

I went up there to check. Here are a few images from my walk there.

I went to this brush complex (location number 1 in the map in figure 15.19, in a spot just off the trail) and took a close-up photo, pointing my camera in the direction of the arrow (and standing at the tip of the red arrow at latitude 37.318673, longitude –122.157772). This is around fifty-six square meters in size and just about the smallest brush complex that I could find with a clear barren zone (figure 15.10).

You can see the barren zone here in the close-up (figure 15.11), while in the satellite photo, it's only visible on the far side of the brush complex. What's interesting here is that the barren zone is hidden in shadows (in the satellite image), but obviously visible at the time of day when I took this picture.

Here's another image (figure 15.12), taken just off another giant brush complex not far away. Look carefully at the image. You can see this one is

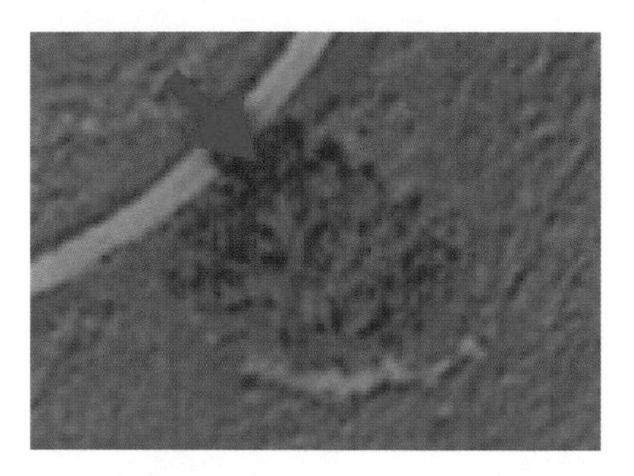

Figure 15.10
Brush complex number 1 from the air.
Image © Google

Figure 15.11
Close-up of brush complex number 1.
Credit: Daniel M. Russell

Figure 15.12
Brush complex number 2 (a large complex of around four hectares in area).
Image © Google

large enough to have trees (the dark spots) growing in the center. (This is brush complex number 2 in the map in figure 15.19.)

As you can see in figure 15.13, I laid out a tape measure so that you can see the width of the barren zone (just over thirty-nine inches or one meter). This zone is so wide that it's also used as a deer trail (with lots of hoofprints making a clear deer path here). For some of these barren zones, it seems that there's more than just herbivory by small mammals going on, at least in the larger brush complexes.

This close-up (figure 15.14) was taken on the west side of the large brush complex. Given the mature size of the brush (a mix of coyote brush chamise and even a few small oak trees in the middle), I'm not surprised that this brush complex shelters more than just rabbits and ground squirrels (figure 15.15).

And another photo—this time a smaller complex. Brush complex number 3, shown in figure 15.16, is around eighty-one square meters.

In figure 15.17, you can see that the barren zone is a little smaller than those of other complexes. This leads me to wonder if the width of the barren zone depends on the area of the brush complex.

After looking at a LOT of brush complexes, I noticed that at some point, the brush complex gets too small to even shelter small mammals, and as a consequence, do NOT have any barren zones, as in figure 15.18.

Figure 15.13
A tape measure to show the size of the gap.
Credit: Daniel M. Russell

Figure 15.14
The above overview was taken here, on the west side of brush complex number 2.
Image © Google

Figure 15.15
This is the close-up of brush complex number 2.
Credit: Daniel M. Russell

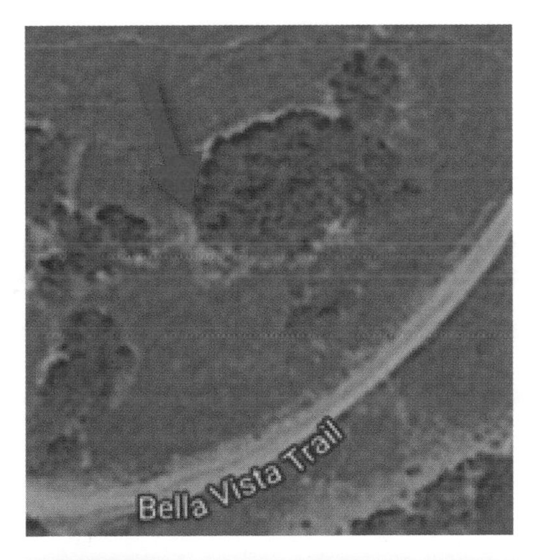

Figure 15.16
An overview from Google Earth showing brush complex number 3.
Image © Google

Figure 15.17
Another tape measure showing the smaller gap next to brush complex number 4. Credit: Daniel M. Russell

Here's an overview of all five clumps of brush (figure 15.19).

Looking through all the documents that we found, one of them—the Point Reyes National Seashore and North District of Golden Gate National Recreation Area, *Fire Management Plan: Environmental Impact Statement* (2004)—leads directly to the "animals in the bush are causing the bare zone" interpretation. The document notes that "shrub cover subsequently increases numbers of rabbits and small mammals (and birds too) that reduce herbaceous vegetation and favor shrub development. Thus, well-established coyote brush stands generally have depauperate understories."

(You have to look up "depauperate." I did. It means an ecosystem that's lacking in numbers of species, or a kind of ecology that isn't diverse.)

But it's natural to seize on the allelopathic effect for chamise, probably because that's the first explanation that pops to the front when you do the obvious search. As you read carefully, however, you'll see that this might be true only for chamise, and doesn't hold for coyote brush and other plants in the area.

Figure 15.18
A small coyote bush with no barren zone (too small). Brush complex number 5.
Credit: Daniel M. Russell

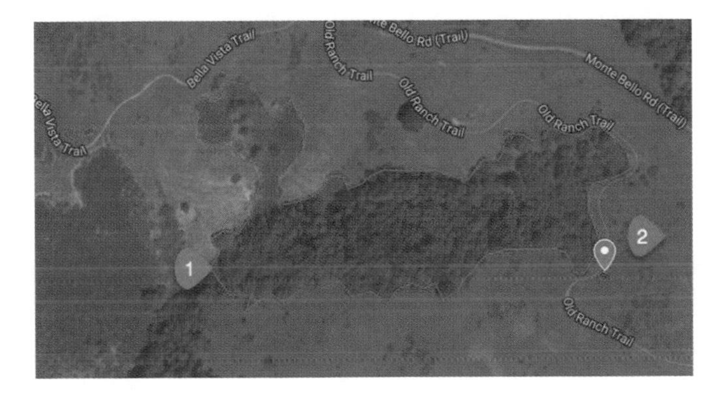

Figure 15.19
Here you can see all the brush complexes. The blue pin marks brush complex number 1, and as you can see, it's relatively small. Another blue pin marks brush complex number 2.
Image © Google; edited by Daniel M. Russell

I suspect that what ALSO happened is a kind of natural reaction when you're searching; we saw the first explanation (allelopathy) and think "that must be it." But as a good researcher, you have to resist that temptation. Just because you saw it first doesn't mean anything. The biggest problem with finding something that seems to fit your data is that you tend to see everything else through that lens; everything you see afterward seems to agree with that initial finding.

If that happens to you, then it's an example of **confirmation bias** at work. This is the tendency (when doing research) to *confirm* what you already believe, latching on to a plausible explanation and then searching for reinforcing evidence, rather than looking for something that might dispute the claim.

Research Lessons

1. *Be careful about confirmation bias!* It's easy to search for information on a topic and find articles that tell you—that is, *confirm*—what you ready know. Be careful and look for a different point of view. In particular, I recommend that you …

2. *Look for another explanation, or at least another point of view.* In this case, it turned out that allelopathy IS a real thing, but it's probably not what's going on here. I found the "animals cause the bare zone" story because I noticed that these articles were all a few years old. So I looked around a bit more and discovered rather-different explanations for what was going on.

3. *It's a good idea to find a recent reference to see if the current belief is the same as older beliefs.* And in this case, they weren't. Botanical thinking had changed since the first allelopathic explanations were written.

4. *You can do a lot of research by using multiple sources to get different takes on the same underlying content.* The big surprise was that so much of this initial work can be done through satellite photos. The barren zones (at least of the larger brush complexes) is quite visible on Google Maps or Google Earth.

5. *Sometimes you still have to get out into the world to check.* Of course, even after doing all the online research you can, there are times when you still need to go out into the field and validate, but I'm constantly impressed by how much early phase science can be done remotely. You can sit at home doing online search and discover entirely remarkable things.

My wanderings on Black Mountain aren't really a substitute for a proper field study. To do real, proper, careful science, I'd have to do a lot of field measurements and careful tabulation of my findings.

But I have to say that I looked at a LOT of brush complexes up there (probably a hundred or so), and in NO case did I find a barren zone around a brush complex that was smaller than ten square meters. If the allelopathy hypothesis were correct, then you'd expect to see a tiny barren zone—but not nothing.

On the other hand, I did notice a few large brush complexes that didn't have a barren zone—but in all cases, they were on fairly steep slopes. I hadn't anticipated that, although it fits in with the small mammal herbivory effect causing the barren zones. Squirrels don't seem to mind a slope, but I suspect that rabbits don't want to be grazing downslope of a brush complex if a predator comes along. Running uphill isn't a great defensive strategy.

Try This Yourself

As you look around the world (say, on Google Earth or another planetary image system), you'll see some fairly remarkable things—patterns and features that cry out for an explanation. Here are two that are worth a bit of research.

1. What are these vertical stripes in figure 15.20? They're located in Florida at 25.3925583, –80.3561022. Take a look and see if you can figure out why they're there. (And once you understand that, maybe you'll discover what particular kind of large animal is attracted to them!)

2. This is an odd Earth pattern (in figure 15.21) that asks for some research. This strikingly circular lake is in Canada at 61.276851, –73.6827605 about 1,000 miles (or 1.6 kilometers) north of Quebec City. It also has no streams coming in or out—yet there are fish in the lake. Can you determine what the story of this remarkably round lake is?

As you do your research, be aware of what your biases are and what you need to do in order to not get stuck on your first impression. As we learned in this chapter, you will probably need to look at multiple resources to get a complete understanding of what's going on in each of these images.

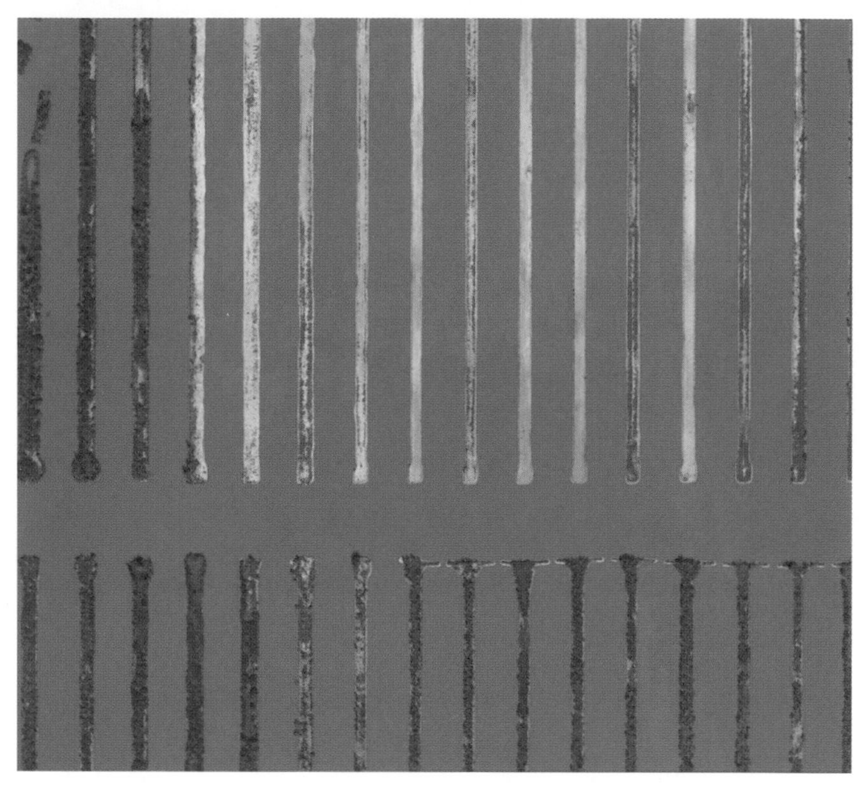

Figure 15.20
This unusual pattern is caused by something large and planned. Can you figure out what this is?
Image © Google; edited by Daniel M. Russell

Figure 15.21
A perfectly circular lake seems artificial, and with no inflow or outflow, it's an amazing body of water. What is it? What kind of fish would live in such a place?
Image © Google; edited by Daniel M. Russell

16 Is Abyssinia the Same as Eritrea? How to Find Additional Context Information for Your Research

Not only do names for ordinary things change, but just as important, people and countries change their names. This happens a lot. But tracking the changes in the names of countries is a somewhat-different skill set than understanding why we rarely talk about people with rose catarrh any more.

Annoyingly, finding names is often simple, but it can get complicated quickly if (and when) names change—whether it's the name of a person or country.

People change names, frequently when they marry or for professional reasons. Take, for instance, Gordon Matthew Thomas Sumner, aka the rocker Sting. Names can get arbitrarily complicated when people change their professional names multiple times (such as the rapper born as Calvin Cordozar Broadus Jr., who also has been known by the names Snoop Rock, Snoop Doggy Dogg, Snoop Lion, DJ Snoopadelic, Snoopzilla Bigg, Snoop Dogg, and Snoop Scorsese).

Even quite ordinary people have variable names. I'm "Daniel Russell" in some, slightly formal places. At the same time, I use "Dan Russell" in situations that are pedestrian, and "Daniel M. Russell" in my professional writings. Of course, I'm "Daniel M. Russell, PhD" when I need to be (such as in formal settings, like professional visits to Japan, where having a professional credential attached to your name gives you a bit of cachet, or whatever the Japanese equivalent of *cachet* is).

And to make it REALLY difficult, there are artists like Prince Rogers Nelson who have a surfeit of names. Prince Rogers Nelson, aka the single-name artist Prince, also used the songwriter aliases of Jamie Starr, Joey Coco, Tora Tora, Alexander Nevermind, and Christopher Tracy. He is, of course,

Figure 16.1
Go ahead and think about how to search for this artist's symbol/glyph/name. The good news is that using the **Search by Image**[A] method works pretty well here.
Credit: Estate of Prince Rogers Nelson

also referred to as The Artist Formerly Known as Prince (TAFKAP), with a difficult-to-search-for symbol (figure 16.1).

Names can be difficult to search for—both names of people (or symbols of people) and places. While you might think that tracking the changes in names of places (cities, countries, and neighborhoods) is similar to the way that one tracks the changes in people names, the fact that it's about geographic names puts a whole new spin on the topic.

This question came up while I was reading a book about the history of Venice. To us English speakers, the place we call the city of Venice is known as Venezia in Italian, but it also has a nickname, La Serenissima, which is the poetic, graceful, affectionate, and diminutive form. But historically it was known to Latin speakers as Venetica, and as Ἐνετοί to the Greeks. Which of these is the city's true name? Of course, they're all right; it just depends on who you're talking to, in what language, and when. We don't often think about time as a determiner of names, but it's an important thing to remember when you're doing online research.

For instance, the Greek island of Crete, with its long history, has many names: in modern Greek it's Κρήτη ('kriti), in ancient Greek it's Κρήτη (Krḗtē), to the Venetians it's Candia, and before that it was known to the ancient world as Κρήτη (Kaptara)! None of those are spelled "Crete," the way I'd search for it in English.

Figure 16.2
The Blue Mosque in Istanbul (not Constantinople)
Credit: Daniel M. Russell

This is an important issue when you're searching for information about a place. If you're lucky, the articles that you're searching for will have all those names listed somewhere. But you can't always count on it, especially when you want to start looking for historical references and content.

Probably the best-known name-shifting city is the city currently known as Istanbul, which was previously known as Constantinople. (There's even a song about the name change! Written in 1953, this novelty song, with lyrics by Jimmy Kennedy and music by Nat Simon, titled "Istanbul (Not Constantinople)," also points out that "Old New York was once New Amsterdam," which makes the point lyrically.)

Since Venice is such an important place in European history, I've been exposed to many of the variants of its name. But oddly enough, I grew up in Los Angeles, which you'll sometimes hear called the City of Angels, even though the hometown baseball team is called the Dodgers because it was an immigrant to the town in the 1950s. The Angels baseball team is from

Anaheim, about sixty miles southwest of downtown Los Angeles. In any case, you'd think that I'd know about the variations of the city's name. So it was with some surprise that I read about the connection between Venice and a city known as El Pueblo de Nuestra Señora la Reina de los Ángeles de Porciúncula. From the context and name, I thought it might be Los Angeles as I know it, but reading this led to the following research questions.

Research Question 1: *Where is El Pueblo de Nuestra Señora la Reina de los Ángeles de Porciúncula? What is that city called today?*

Research Question 2: *Because the book I was reading was set in the 1700s, it was rich with historical references and so made me wonder, What would that place (El Pueblo ...) have been called in 1600, before the Spanish arrived?*

I started by just searching for the long form of the city name, mostly because it's SO specific that nothing else will match this magnificently wordy place-name:

[El Pueblo de Nuestra Señora la Reina de los Ángeles de Porciúncula]

That search tells you quickly that this is the place also known as Los Angeles, California (aka LA). The Wikipedia page shares the story that Franciscan monk Juan Crespí tells about the Portolá expedition being impressed by a river it named El Río de Nuestra Señora la Reina de los Ángeles de Porciúncula—meaning "the River of Our Lady Queen of the Angels of Porciuncula." The name derives from Santa Maria degli Angeli, a small town in Italy housing the Porciuncula, a tiny plot of land that held the church where Saint Francis of Assisi lived. Various versions of Crespí's name would be used for the town, including the exceedingly long El Pueblo de Nuestra Señora Reina de los Ángeles sobre el Río Porciúncula and other variations on that theme.

Other variations that I've seen in my research (from a variety of sources) include:

El Pueblo de la Reina de los Angeles Sobre el Rio de Porciúncula

El Pueblo de Nuestra Señora de los Angeles de la Porciúncula

El Pueblo de Nuestra Señora de los Angeles

El Pueblo de Nuestra Señora la Reyna de los Angeles

El Pueblo de Nuestra Señora la Reyna de los Angeles del Rio Porciúncula

Pueblo del Rio de Nuestra Señora la Reyna de los Angeles de Porciúncula

Of course, Gaspar de Portolá's expedition didn't reach the Los Angeles Basin until 1769. What would the locals have called that place before he arrived?

As you know, Los Angeles is a huge place—so when I think about "that place," what do I mean? The point where the city was founded? The largest nearby settlement of Indians who live nearby? Where the first building was constructed? How about the first Catholic mission?

How would you find out? Here's my general query:

[History of Los Angeles]

It leads to the Wikipedia article on Los Angeles history, which has a section on "prehistory," which tells us that the Tongva people who inhabited the area when the Spanish arrived called the Los Angeles region Yaanga in their language. And what's more, the Spanish pueblo was located near a large village on the river.

To complicate things, the DiscoverLosAngeles.com website claims that around AD 500, Tongva Indians settled in the Los Angeles Basin, displacing the previous inhabitants, the Tataviam (later called the Fernandeno).[1] By the sixteenth century, the region's main village was called Yang-Na, near present-day Los Angeles City Hall.

We've got two somewhat-different names for the pre-Spanish village. Let's do some background checking here. What sources are cited for each of the names?

Yaanga is cited in the Wikipedia article in a book by Pamela Munro and colleagues, titled *Yaara' Shiraaw'ax 'Eyooshiraaw'a. Now You're Speaking Our Language: Gabrielino/Tongva/Fernandeño.*[2] When I search for this book on Lulu.com (a self-publishing site), however, I find that there are two books with different authors. One is by Pamela Munro, and the other is by Julia Bogany. This is slightly suspicious; it's a self-publishing site. BUT when I search for these authors by name, I find that Munro is a linguist at the University of California at Los Angeles with extensive work in Native American languages, including Tongva! Bogany turns out to be a Tongva elder who teaches the language and culture of the Tongva. These aren't just random self-publishers with a passing interest; they are world experts on the Tongva language.

If you're ever going to find two experts in the Tongva language, these are the people you'd find. And when you look at both books, it's clearly the same book; one is just a "large print" edition.

I was able to find Munro's email address without much trouble—so I wrote to her and asked her opinion about the name. She graciously wrote back (almost instantly!), saying, "The basic form of the name was probably Yaar or Yaay; we don't know. Different endings are added to the root Yaa- (not a word). The -nga ending means 'in' and is also used for the form people usually give for village names."

How about the other name for the village?

Yang-Na is used in the DiscoverLosAngeles article, but has no citation for the source of the name. In this case, I did a quoted search:

["yang-na"]

I discovered a lot of Chinese artist results and a mention in US history (again, without any reference for the name), but a mention that Juan Rodriguez Cabrillo came across the Yang-na village in 1542, noting the location on his map as he continued his exploration.

I shifted my search to:

["yang-na" Cabrillo]

That search then gave me a bunch of results. This name is used just about everywhere. But after extensive searching, I haven't been able to find a decent citation about where the name came from OR Cabrillo's map with the name "Yang-Na" on it. (I'll keep looking, but right now, it seems like a less used variation of the other name.)

So let's go with what the Tongva experts (and Wikipedia) call it: **Yaanga**.

Of course, this kind of research lends itself to a bit of spreading out of interests. As I was reading about El Pueblo and Yang-Na, I ran across a lot of other place-names. Having grown up in Los Angeles, I knew many of them, but I kept seeing a reference to a place called **Humqaq**. That's an unusual name, yet it was clearly local to the greater Los Angeles region. So what's the story behind **Humqaq**?

The obvious query [Humqaq] tells us that this is Point Conception, just west of the Southern California city of Santa Barbara, northwest of Los Angeles by about a three-hour drive up Highway 101.

The Wikipedia article about Humqaq (aka Point Conception) says that "Point Conception was first noted by Spanish maritime explorer Juan Rodriguez Cabrillo in 1542 and named **Cabo de Galera**. In 1602, Sebastian Vizcaíno sailed past again, renaming the protruding headland **Punta de la**

Figure 16.3
Point Conception on Google Earth. Santa Barbara (and Los Angeles) are to the east. Image © Google

Limpia Concepción ('Point of the Immaculate Conception'). Vizcaíno's name stuck, and was later anglicized to today's version."[3]

If you keep reading, you'll eventually find out that "it was called Humqaq ('The Raven Comes') in the Chumashan languages."[4] So the name *Humqaq* is the Chumash name of the point.

After the lesson of Yaanga, it's worth doing a bit of checking. To follow up and find a second source, I used this information in a query:

[Point Conception Cabrillo]

This finds multiple confirming sources. For instance, it found Paul A. Myers's 2004 book, *North to California: The Spanish Voyages of Discovery, 1533–1603.*[5] This book tells exactly the same story, giving great, high-quality references that you can check.

The trouble with reading history is that the questions just keep coming. Remember that the book I was reading was set in the 1700s and it kept mentioning a place called Óbuda. From the context, I knew it was in southeastern Europe, not so far from Venice. So naturally this led to my next question.

Research Question 3: *Where is the city of Óbuda? And what's this place called now?*

As I did last time, I started by simply searching for the city's name. (Notice that I didn't try a complicated query at first; strategically, I usually begin my searching with a short, simple query. That way I pick up on any ambiguous uses of the terms and don't get sent down a rathole of spurious results. Start simple; it generally works out for the best.)

[Óbuda]

Any number of resources linked by this search tells us that this is or was a city in Hungary that was merged with Buda and Pest in 1873. This area now forms part of District III-Óbuda-Békásmegyer of Budapest. The name means Old Buda in Hungarian (or in German, Alt-Ofen). The name in Croatian and Serbian for this city is Stari Budim, but the local Croat minority calls it Obuda (the name Budim is used for the fortress in Buda).

So it's Budapest. Again, to check, I did a query for:

[history Budapest]

I found Budapest's official city website, which tells the same story.[6]

Now that we're thinking about geographic name changes, What other places have interesting name shifts over time?

With its turbulent history, Africa is full of name changes (and even official language changes) over time. Here's another research question about a country that's undergone a number of changes with time.

Research Question 4: *What was the historical name of the country where the city of Dar Es Salaam is today—say, pre-1964?*

To answer this question, you're going to need to start looking for historical documents about the city of Dar Es Salaam. That's fine, but do you know off the top of your head what country you need to look for? I didn't, so I began finding out a bit about the city of Dar Es Salaam with this query:

[Dar Es Salaam]

I learned that it's in Tanzania. That's great, so the next obvious query here is:

[history of Tanzania]

I found Google's web answer: "*On 26 April 1964, Tanganyika united with Zanzibar to form the United Republic of Tanganyika and Zanzibar. The country was renamed the United Republic of Tanzania on 29 October of that year. The name Tanzania is a blend of Tanganyika and Zanzibar and previously had no significance.*"[7]

I have to admit that I grew up with Tanganyika as a country in Africa and was vaguely aware that it had ceased to be sometime in the mid-1960s, but I didn't know that the new country name is a portmanteau word combining Zanzibar and Tanganyika!

So if you're searching for information about the region around Dar Es Salaam, you'd want to know about this name change. Sometimes you have to look not just for the answer to your research question but also for the answer to a larger, more encompassing question—in this case, the history of the country.

Speaking of countries in Africa, the shifts in political fortunes sometimes makes asking questions a little tricky. Be careful about what you ask, as you just might have hidden assumptions built into your question. For example, here's a reasonable-sounding research question.

Research Question 5: *What was the name of the capital of Zaire in 1900?*

Again, here's another African country name that I was aware had changed, but I didn't really know much about the transformation process. This is the kind of research problem that has to take the passage of time and potential changes into account.

The obvious search is:

[capital of Zaire]

It gives you the following result on Google (figure 16.4).

But be careful! While this result is true, if you just accept this answer straightaway, you'd be missing an important fact. In 2018 as I write this book, Zaire is no longer a country. It's always a good idea to do a quick double check about the surrounding information, especially if you know that African countries have a tendency to shift names in the course of history. So if you now do a background search to learn about Zaire, you'd do this:

[Zaire]

Figure 16.4
A search for [capital of Zaire] shows Kinshasa as the result. Be sure to check the context! Credit: Google and the Google logo are registered trademarks of Google Inc., used with permission

You'll quickly find out that Zaire was the name for the Democratic Republic of the Congo—a country that existed only for twenty-six years, between 1971 and 1997.

This is a crucial point: check your work! Be sure you get the context surrounding what you're searching for! Kinshasa IS now the capital of the Democratic Republic of the Congo and was the capital of Zaire.

Yet remember the research question? "What was the capital of Zaire in 1900?" This is a bit of a tricky question, but it's the kind of question that comes up all the time. It's the type of question that arises in real research; search questions don't necessarily have to make sense or be internally coherent. That's the way this question is: Zaire didn't exist before 1971. So asking, "What's the capital in 1900?" is an inherently odd question.

What, then, is a reasonable interpretation?

I'd say the question is really like this: What WAS that part of Central Africa called in 1900?

Reading about the history of Zaire tells us that it was formed out of the Democratic Republic of the Congo in 1960.[8] Before 1960, this country was called the Belgian Congo, which was established by Belgium through annexation in 1908.

OK, what was it before 1908?

Reading carefully, you'll find that a group of European nations (the Berlin Conference), divided up the area of the Congo, ceding it to Belgium. As a result, King Leopold II of Belgium received a large share of territory (2,344,000 square kilometers, or 905,000 square miles) to be organized as the Congo Free State. The Congo Free State operated as a corporate state privately controlled by Leopold II through a nongovernmental organization, the Association Internationale Africaine. The state included the entire area of the present Democratic Republic of the Congo (and therefore Zaire) and existed from 1885 to 1908, when the government of Belgium annexed the area.

So what was the capital of the Congo Free State?

[capital Congo Free State]

We find that Boma was the capital city of the Congo Free State and Belgian Congo (the modern Democratic Republic of the Congo) from 1886 to 1926, when the capital was moved to Léopoldville (since renamed Kinshasa).

To double check, I search:

[history of Boma Congo]

That leads to several books, such as the *Historical Dictionary of the Democratic Republic of the Congo*, containing background information that agrees with this.[9]

I would say that in 1900, Boma was the capital of the region of the Congo that was known as Zaire and is now known as the Democratic Republic of the Congo.

Interestingly, as I was doing this research, I tried other search engines as well. (You should too, from time to time, just to be sure you know what is possible.) The results are variable. Sometimes they're right, and sometimes they're almost yet not quite right. More often, they present the information without the surrounding context that you need to really understand things.

On the other hand, there are other search engines that get it definitively wrong, telling us that the capital is Kinshasa, which can't be right. (The Democratic Republic of Congo didn't exist in 1900.)

Figure 16.5
Credit: Google and the Google logo are registered trademarks of Google Inc., used with permission

Google doesn't give all the details, but just gives ordinary search results (which is probably really the right thing to do—although see chapter 20, since in the future, this will probably change). This is an important skill for researchers; don't overread what's in the result, but see it for what it is. In this case, you must go deeper to see the context of Zaire in 1900.

Since this is true, is there a single reference work (hopefully online!) that would let you answer all these questions? What would a work be called?

One kind of useful document to know is the *gazetteer*. It's basically a geographic dictionary or directory that has information about the geographic makeup as well as social and physical features of a country, region, or other place. A gazetteer usually has the subject's place in the world and almanac-style information (dimensions of mountains, rivers, oceans, waterways, population, gross domestic product, literacy rate, and so on).

A *historical gazetteer* is a gazetteer with information about place-names and events, and how those change over time. To find such things, you first have to know what a "gazetteer" is and then you can search for historical versions of the same with this query:

[historical gazetteers]

It finds several pages of gazetteers and collections of maps. I found that the National Geospatial Intelligence Agency's mapping gazetteer works well to find place-names (and associated data) from all around the world.[10]

Los Angeles
Los Angeles County *Southwestern California, on Pacific Ocean*
1542 Spanish explorer Juan Rodriguez Cabrillo sails Los Angeles Harbor in his caravel, names it Bay of Smokes and Fires for the numerous Native American camp fires; **1769** Native American village of Yang-na discovered here by Capt. Gaspar de Portolá; **Sept. 1771** Mission San Gabriel founded; **Sept. 4, 1781** pueblo founded by Father Junipero Serra and Don Felipe de Neve, governor of Spanish California, with 44 settlers at Yang-na, named pueblo Nuestra Senora la Reina de Los Angeles de Porciuncula (Our Lady of the Queen of Angels of Porciuncula); **1805** American ship *Lelia Byrd* anchors at San Pedro; **1817** capital of California Territory alternates between Los Angeles and Monterey; **1818** Joseph (José) Chapman becomes first English-speaking settler; **1822** Church of Our Lady the Queen of the Angels completed; **1840** Lugo House built by Don Vicente Lugo; **March 9, 1842** gold discovered in Placerita Canyon, north of San Fernando Valley; **1846** California seized by U.S.; Commodore R.F.Stockton captures town in bloodless battle; **1850** Los Angeles County established; incorporated as a city;

Figure 16.6
A preview from the *Historical Gazetteer of the United States* showing the full name of Los Angeles, with the previous name as well.
Credit: Taylor and Francis Group LLC, Books

Using this query, a little further down on the page I found the *Historical Gazetteer of the United States* in Google Books (by Paul T. Hellmann) in which I found the following entry for Los Angeles (by using the search box on the Google Books page).[11]

Note the highlighted passages. This reference book uses the "Yang-na" form of the Tongva village name and gives one of the variations on the long Spanish name of Los Angeles.

Another useful reference to know about is the "dictionary of place-names" that several reference publishers issue. It's often useful to search for such a dictionary for a given place. Here are a couple of examples that I've used:

[dictionary place-names]
[dictionary place-name British]
[dictionary place-names California]

These will find resources that can answer all kinds of questions about how geographic place-names change over time!

Finally, the last research question on place-name changes.

Research Question 6: *IS Abyssinia the same as Eretria?*

First, exactly what is Abyssinia? Let's do a quick overview search:

[Abyssinia]

We can easily learn that Abyssinia was "the Ethiopian Empire, historically known as Abyssinia, a nation that comprised the northern half of present-day Ethiopia." But as we know, you should read the entire description, and thus you'll discover on the Wikipedia entry for Abyssinia that "Ethiopia, the modern nation remains known by its (name) Abyssinia."[12]

Clearly, the history is a bit tangled here.

Research Lessons

There are a few lessons.

1. *Check your web answers!* As you can see, various sources give variations on a theme. As I always say, double source (or more!) your answers. Get your information from places you can trust.

2. *Remember that you can always ask a real expert.* In this case, Munro was a fantastic resource; she's a world expert on the Tongva language, but was gracious in giving me an answer to a fairly technical question. People are generous, as a rule. Reach out to them.

3. *Be sure you understand the question.* In the Zaire capital question, the answer took some digging to get to the bottom of the issue. It really was NOT obvious what the answer was and took some thinking about what a possible answer would be. Answering the Zaire question also meant that we had to realize that the question had a built-in assumption—that is, that Zaire was a currently existing country. You should check your web answers, but also be sure to check your assumptions.

4. *Sometimes questions take a while to answer.* To this day (late 2018), I'm still looking for the Cabrillo map with Yang-na on it. (Or something similar.) It may take a while, but you never know; I'm constantly on the lookout.

5. *Know what a gazetteer and dictionary of place-names can do for you.* Part of being a great online researcher is knowing that certain kinds of resources exist. Just as you should know that a collection of maps is an *atlas*, it's worth knowing that gazetteers and place-name dictionaries exist. Not only do they pull together a particular kind of information, but they often have high-quality, credible information on related places and topics that will prove useful.[13]

6. *Get context around what you find!* Can you answer those little questions, such as where, when, and how? What other relevant things were going on

at the time? Who are other players in that time and place? Reading a Wikipedia article is often a great way to get started in picking up the context around a topic, but consider other contextualizing sources as well (overview articles, summaries of a topic, etc.).

Bear in mind that what I'm showing you here is what's true NOW. As more books and articles are published on the origins of these places, the stories could—and most likely will—change.

How to Do It

A. Search by image. Suppose you have an image of something—a building, logo, or distinctive kind of leaf. You can search for that *image* over all the images that Google knows about.

As an example, here's an image of a beautiful leaf that I found on the ground last fall.

I opened up a Google Images search (you can use the URL Images.Google.com).

Figure 16.7
I took a photo of the leaf in question on a plain background. (You don't always need a blank background, but for leaves, it's best to have it in isolation.)
Credit: Daniel M. Russell

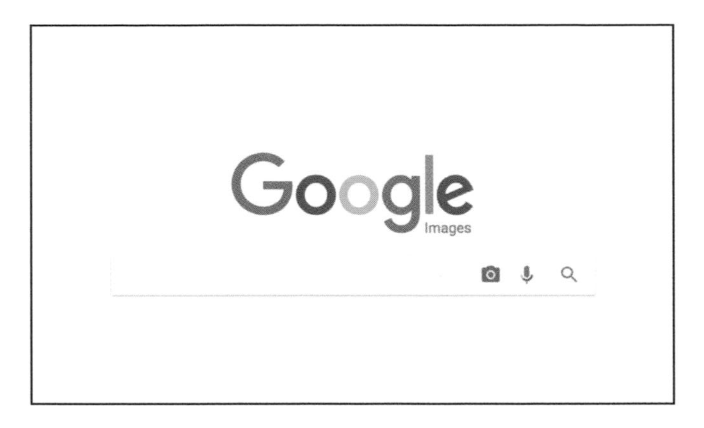

Figure 16.8
A Google Images search can accept other images as input. You can drag an image from your desktop onto the search bar, or click on the camera icon to give it a URL or upload an image.
Credit: Google and the Google logo are registered trademarks of Google Inc., used with permission

When you see the camera icon in the image search, you can click on it and upload an image (like the leaf image above), which then does a search for that image, giving results with images that are near matches. In this case, we discover that this leaf is from a ginkgo tree (aka Maidenhair tree; see figure 16.9).

Notice that the name of the image file (MysteriousLeaf.jpg) is shown in the search bar (highlighted in blue) next to "maidenhair tree," which is Google's best guess at what the most plausible textual query for this might be.

Try This Yourself

Not only do people and places change their names, but they change their identities, sometimes significantly. Austria, for instance, was the basis for the Austro-Hungarian Empire from 1867 until 1918. Its border has changed more than once or twice in the past three hundred years. This kind of shape-shifting is frequently accompanied by name shifting as well.

As an example, can you work out the multiple names, shapes, and sizes of the country currently known as Tanzania? It's in the southeast of Africa, in a place often called the Swahili Coast.

What resources would you need to use to find out the historical changes? (Remember that sometimes the least likely resources can prove useful!)

Figure 16.9
A search by image query takes an image and searches for similar images as well as gives information about the most plausible search query—in this case, this is a leaf from a ginkgo tree.
Credit: Google and the Google logo are registered trademarks of Google Inc., used with permission

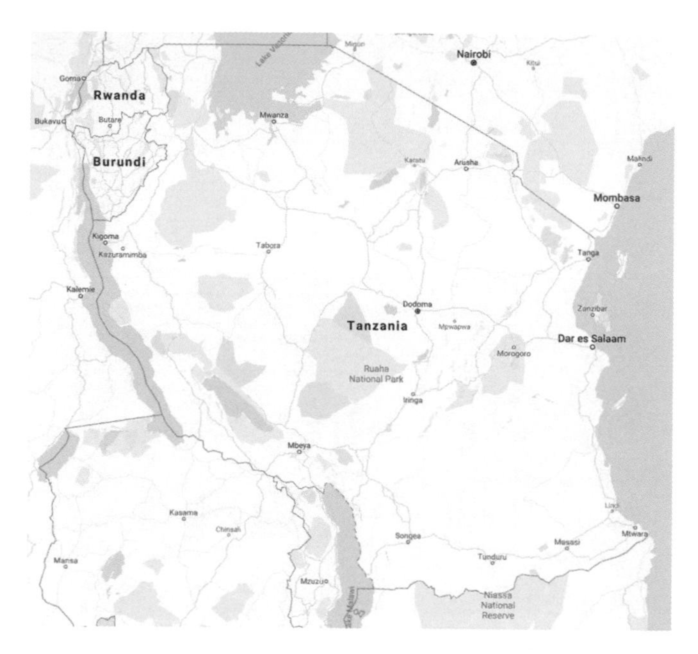

Figure 16.10

The place in Africa now known as Tanzania has had several name changes over the past two hundred years. Can you find out what other names (and country shapes) it has had?

Credit: Map data © Google

17 The Mystery of the Parrotfish, or Where Does That White Sand Really Come From? How to Triangulate Multiple Sources to Find a Definitive Answer

I mean, is that beautiful sandy Caribbean beach really something less savory? How to find out where sand comes from by using online research methods and triangulating across multiple information sources. You'll find out something deeply surprising.

It's the beginning of September, and it's just about the perfect day. I'm spending a few days on the small Dutch Caribbean island of Bonaire, scuba diving in the gentle waters just off Kralendijk. It's like I'm living in one of those PBS adventure TV specials—the water is 82 degrees and ultraclear, and full-of-life seas surround the island.

When you scuba dive in the Caribbean, you're constantly seeing things that you don't know, and consequently, spending a fair bit of time searching for answers to "What kind of fish was that?" or "What IS that strange thing we found that looks like a snake, but has gills?" This happens pretty much every time you go underwater, and for curious people, it's a place of endless entertainment.

This is a fish you see everywhere around Bonaire (and most of the Caribbean as well). I shot this picture in the afternoon at a depth of 30 feet (or 10 meters). This fish is about 2 feet long (or 0.6 meters) from the tip of its tail to its nose. This is a typical pose as it swims on the reef. These creatures bump around the coral reefs, apparently biting the reef every minute or so, wandering from place to place. And if you watch one for more than a couple of minutes, you'll see that in mid swim, it will suddenly excrete a long, fine cloud of what looks like white dust and tiny pebbles, streaming out into the clear water from someplace near the bottom of its belly.

Figure 17.1
A common fish in the waters around Bonaire.
Credit: Randall Spangler

Figure 17.2
This is where sand starts, with a parrotfish approaching the coral and giving it a good bite.
Credit: Daniel M. Russell

Figure 17.3
That moment when sand is born from the bottom of a parrotfish.
Credit: Randall Spangler

Here's the magical moment: a parrotfish swims by and a stream of sand comes from the bottom of the fish.

When I come up to the surface after watching this for a bit, I can't help but wonder a couple of basic things.

Research Question 1: *What kind of fish is this? What's both the common name and Latin scientific name?*

Research Question 2: *What's going on when this fish is excreting that cloud? (I'm wondering because it happens SO often. I never see any other kind of fish doing this. Is this fish odd somehow?) How much sand does this parrotfish poop in a day?*

I started with the basics: What is this beast? I began by searching for:

[Caribbean fish guide]

Why this? Because I was hoping to find some kind of authoritative reference to the fish of the Caribbean and maybe pick up some language that would help me describe what kind of fish it is.

Figure 17.4
Reefguide.org's parrotfish page, including our fish, the stoplight parrotfish.
Credit: Florent Charpin, Reefguide.org

In the search results, I found Reefguide.org, which has a high-quality guide to the reef fish of the world.[1] It has sections for tropical reef fish by location (Bahamas, Caribbean, South Pacific, etc.) and type (parrotfish, angelfish, goby, etc.).

From its home page, I chose "Caribbean, Bahamas, Florida" and then did a quick scan of the fish families. Just looking around tells you a lot; you learn both the range of different fish types *and* some of the language that's used to describe a fish. It wasn't a "small oval" fish or any of the others I saw, so I kept on looking (the web equivalent of turning pages) until I found the **parrotfish** category. Once you see the parrotfish page, it's pretty clearly that kind of beast. The fish in my picture is a dead ringer for the one that's in the top row, third from the left.

Just to double check that this is in fact the same animal, I did a regular Google search for its name:

[stoplight parrotfish]

That search confirmed that it really does look like this fish. I also picked up that this fish is called a *Sparisoma viride* by scientists. Knowing a technical term like this will come in handy later. Why? Because ordinary names like

Figure 17.5
A juvenile stoplight parrotfish looks different from the adult form. This is true for an amazing number of fish; they're not just smaller versions of the adults but also colored and shaped rather differently.
Credit: Daniel M. Russell

stoplight parrotfish too often refer to different fish in different places. But a Latin name like *Sparisoma viride* will only ever mean this one particular kind of fish (that's the promise behind the biologist terminology).

I also found out, remarkably enough, that male parrotfish keep harems of females nearby. When the dominant male dies, one of the females will change gender and color, turning into the alpha male for the harem.

There's also a striking difference in appearance between the adults and juveniles. Here's a photo I took of a juvenile stoplight parrotfish.

I was *really* curious about the effects of all that excretion. You see this happening ALL the time. And of course, on the dive boats, you hear stories about how the parrotfish really are responsible for much of the sand you see on the beaches. I needed to learn something about the relationship between parrotfish and geology. I started with the parrotfish I knew about by doing a query like this:

[stoplight parrotfish sand]

The results, though, were too diffuse to be useful.[2] I suspected that adding the word "stoplight" isn't helping my search.

So I modified my query to be:

[parrotfish sand]

I quickly found many articles about the effect of parrotfish (in general, and not just the stoplight parrotfish) on sand creation in the Caribbean.

I soon learned from the Wikipedia article on parrotfish that they'll bite into the coral, scraping off a bit with each bite.[3] That's what I saw them doing. They digest the algae they get in that bite, and then, after a bit of crunching and digestion, the coral bits are excreted into the water, making up a large amount of sand. In essence, much of the white sand on Caribbean island beaches is made up of fish poop! (OK, to make you feel better, it's made up of the excreted, indigestible bits of pulverized coral.)

And in the popular magazine *Newsweek*, I find that some people believe that there can be anywhere between 840 pounds for a Hawaiian parrotfish and 11,000 pounds for some (unnamed) fish.[4] The *Newsweek* article points to an article in the journal *Marine Biology*. (The March 2010 article is titled "Bioerosion of Coral Reefs by Two Hawaiian Parrotfishes: Species, Size Differences, and Fishery Implications," which pretty clearly is relevant to our research question.)[5]

This kind of discrepancy comes up all the time, and it's the reason that you don't want to single source anything, even simple facts like this.

Just to triangulate what I was reading, I did the same search on Google Scholar to see a few scholarly papers on the topic.

I found a paper with what looked like a reasonably good title: "Rate of Bioerosion by Parrotfish in Barbados Marine Environments." It points out that "stoplight parrotfish were observed to spend 80% of their time feeding on surfaces covered with filamentous algae and 20% on sand. Only *Sparisoma viride* consistently leaves well-defined scars on live coral."[6]

But there's a debate among different sources about how MUCH sand a parrotfish produces each year. If you look at the comments, some say it's 1 ton (2,000 pounds) per year per fish, while others say it's 200 pounds per

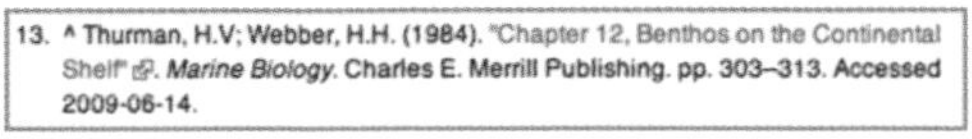

13. ^ Thurman, H.V; Webber, H.H. (1984). "Chapter 12, Benthos on the Continental Shelf". *Marine Biology*. Charles E. Merrill Publishing. pp. 303–313. Accessed 2009-06-14.

Figure 17.6
The crucial citation in the Wikipedia article links to a dead result. Looks like we'll have to get the book and check manually.
Credit: Wikipedia

year per fish. Since that's SUCH a variance, we have to think about the rate of sand production per fish a bit more deeply. If sources vary by that much, something fishy is going on. (Pardon the pun.)

So how MUCH sand does a single parrotfish create in a single year? And where do these numbers come from?

The source for the claim of 200 pounds of sand per fish seems to be from the Wikipedia article on parrotfish, which points to the article by Harold V. Thurman (see below for reference list).

Unfortunately, the link to the Thurman article specified in Wikipedia is dead (it was last checked in 2009), and the target site no longer supports this reference. I spent some time searching for this chapter, and while I can find lots of references to it, I can't find the book chapter itself.

Even worse, many of the references to this chapter I found are clearly copies of the citation and link from Wikipedia. (Even at such well-respected sites as the Encyclopedia of Life, whose parrotfish page would be considered authoritative, merely quote the Thurman reference and give the same link to a dead page.)[7] Sigh. Lesson learned here: references and links seem to rot a bit with time. Which is why you have to constantly check these things.

So I went to my local library's website and interlibrary loan got me the darn thing.[8] Amazingly enough, I got the book after only three days! And there, on page 310, is the assertion: "By eating coral (and corraline red algae) a single parrot fish may produce up to 90 kg (190 lbs) of sand per year."[9]

Case closed? Not yet. Unfortunately and irritatingly, this book doesn't give any reference for that bit of data. Where did it get that information from?

How about the other claim that each parrotfish produces 2,000 pounds per fish per year? The source for the 2,000 pounds seems to originate from the Ichthyology department's page about parrotfish at the Florida Museum of Natural History.[10]

On that page, it says that "parrotfish may produce as much as one ton of coral sand per acre of reef each year."

But read that carefully. This means "all of the parrotfish on the reef in 1 acre" are producing around 2,000 pounds of sand over the entire acre. That's the contribution by all the parrotfish working (so to speak) together.

Sadly, it doesn't give a reference for its source material either. So I search for the exact phrase as seen on the Florida Museum of Natural History site (here I use double quotes around the exact phrase to find *exactly* that phrase):

> ["Parrotfish may produce as much as one ton of coral sand per acre of reef each year"]

It's clear that all the references are using the museum's page as their source. (Typically, much of the text in the article is quoted directly, including that phrase. This kind of thoughtless copying always makes me suspicious.)

By being diligent and following various sources, however, I also found the reference to 2,000 pounds as coming from the book *Fishes of the Bahamas*. The citation would look like this:

Bšhlke, J. E., and C. G. Chaplin. 1993. *Fishes of the Bahamas and Adjacent Tropical Waters*. Austin: University of Texas Press.

So I go to Google Books and find the book, but this citation is slightly wrong. The first author's surname is Böhlke, not Bšhlke. (Some character encoding failed, and people weren't careful about their citations. Sigh. This happens more often than you'd like. Part of your research skill is learning to read through and around such minor errors.)

And once again, this book falls into the "scanned on Google Books, but snippet-view only" (which doesn't help as I can't read the important bits). It's not scanned on Amazon.com, Hathi Trust, or any of the other sources. So back to the library.

I ordered this book on interlibrary loan, and got it about one week later along with the Thurman book. The thing is, *Fishes of the Bahamas* is a giant tome: 771 pages of fish data. There's not much plot in this book, but it's still pretty interesting to read (that is, if you're a diver and/or really love to read infinite details about every possible fish in the sea).

On the other hand, on page 465 of this magnificent tome I found the sentence, "On a Bermuda study reef it was recently estimated that one ton

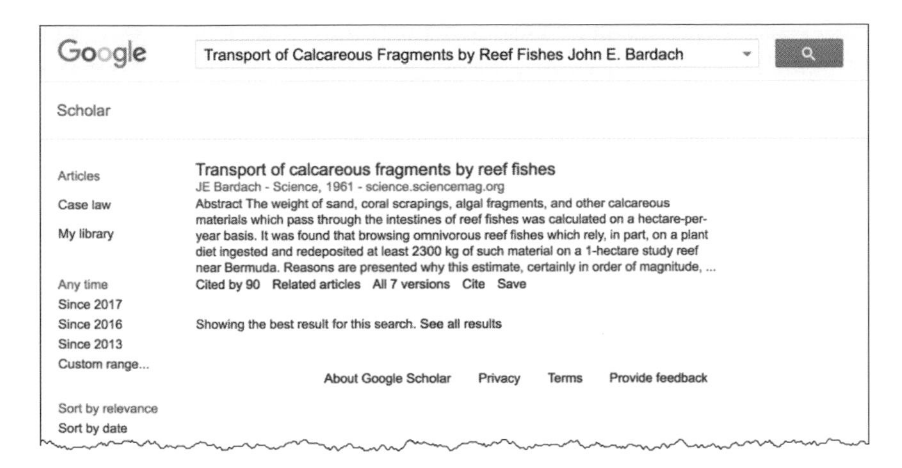

Figure 17.7
Here I just searched on Google Scholar for the title of the paper (which I copied and pasted from that other article.
Credit: Google and the Google logo are registered trademarks of Google Inc., used with permission

per acre per year of this material ["calcareous material," i.e., coral sand] passes through the intestinal tracts of reef fishes and is redeposited as fine sand, and that the *scarids* [parrotfish] are primarily responsible."

And the reference (finally, a real reference!) given here is:

Bardach, J. E. 1961. "Transport of Calcareous Fragments by Reef Fishes." *Science* 133 (3446): 98–99.

By going to Google Scholar and searching for that author plus the title, I was able to get to the original article, "Transport of Calcareous Fragments by Reef Fishes."[11]

Now we're getting somewhere. In that paper (published originally in the well-respected journal *Science*), the author writes, "It was found that browsing omnivorous reef fishes which rely, in part, on a plant diet ingested and redeposited at least 2300 kg of such material on a 1-hectare study reef near Bermuda." I like that word choice: "redeposited." The article means pooped out.

Let's do some Google conversions to check this number.

To convert these numbers, you can do Google queries like this:

[2000 pounds per acre in kg per ha][12]

This query asks Google, "What does 2,000 pounds per acre convert to in *kilograms per* hectare?" The answer is straightforward (and we don't have to do the math): it's 2241.7 kilograms per hectare (a hectare is 10,000 square meters, or just about the size of 2 football fields).

See how that works? Now you can use Google to convert the quantity given in the paper (2,300 kilograms redeposited on 1 hectare") by doing a query like this:

[2300 kg per ha in pounds per acre]

It now tells us how many pounds per acre, which gives you the results of 2,052 pounds per acre. That's pretty close to 1 ton.

I wanted to get the full text of the paper, but that's twenty dollars for a "short-term use" (one-day) license. As you can imagine, that's a little expensive for something that I'm not sure will be what I seek.

But we still don't know how much sand per fish per year that represents. To figure this out we need to know, *How many parrotfish live in on an acre of reef?*

Hoping to figure out this subpuzzle, I started doing searches like this, using the scientific name of the stoplight parrotfish:

["Sparisoma viride" sand OR sediment production]

This is reasonably productive (especially when done in Google Scholar) and leads to several papers on the topic. The most cited paper is:

Bruggemann, J. H., et al. 1996. "Bioerosion and Sediment Ingestion by the Caribbean Parrotfish Scarus vetula and Sparisoma viride: Implications of Fish Size, Feeding Mode and Habitat Use." *Oldendorf* 134 (1): 59–71.[13]

I got a PDF of the paper by doing this search, using the **filetype**: operator to limit my search to ONLY PDF files:

[filetype:pdf "Bioerosion and sediment ingestion by the Caribbean parrotfish Scarus vetula and Sparisoma viride"]

That led me to the PDF file, which I printed out and carefully read.

This is when I realized that this is a much more complex problem than just looking up the number of parrotfish per square meter on the reef.

By reading this (and a few other papers), I learned that the density of parrotfish on the reef varies hugely by depth. (Shallow water has a lot;

water deeper than 50 feet has fewer.) They also sleep at night, and feed only between 7:00 a.m. and 6:00 p.m. (and by the way, they mate daily, between 7:00 a.m. and 9:00 a.m. ... but I digress). I also found out that the quantity of coral ingested when they bite varies a good deal by the size of the fish, but it varies even MORE depending on the kind of parrotfish.

Among other things, we HAVE to be careful not to mix up our parrotfish species (e.g, Hawaiian parrotfish with Caribbean parrotfish, or even parrotfish that live side by side in the waters of Bonaire); as the paper points out, "The erosion rates per bite by *Sparisoma viride* is more than an order of magnitude higher than in Scarus vetula." That is, the stoplight parrotfish (*Sparisoma viride*) has a bite size that's ten times larger than that of the *Scarus vetula* (queen parrotfish).

But we're getting somewhere.

If you read this paper carefully, you'll eventually find figure 5, "*Scarus vetula* and *Sparisoma viride*: Erosion rates per bite (upper graphs) and per day (lower graphs) as a function of fish size." Figure 17.8 is a redrawn version of the relevant chart.

If you look at the graph in the lower-right corner, that's a chart of how big those bites are. Although the grams per day vary greatly by fish size, the average is right around 300 grams per fish per day (marked by the dashed horizontal line). Basic math (which you can do by typing in this expression directly into Google) shows us that:

300 grams * 365 days = 109,500 grams per year

Or 109.5 kilograms per year (204 pounds), which is close enough (given all our averaging) to our earlier estimate of 95 kilograms per year. (And incidentally, it's nothing close to 2,000 pounds per fish per year.)

If this is correct, then to get 2,000 pounds of sand contributed per acre by all the parrotfish, we're going to need around 10 stoplight parrotfish per acre. By my direct observation on the reefs of Bonaire, that seems low, but let's see if we can't find some numbers in the literature.

Another paper I found from my previous query is this:

Choat, J. Howard, et al. 2003. "An Age-Based Demographic Analysis of the Caribbean Stoplight Parrotfish *Sparisoma viride*." *Marine Ecology Progress Series* 246:265–277.[14]

He "conservatively estimates" that there are four individual parrotfish per plots of 300 square meters. OK, but how big is an acre in square meters? Another Google query handles that:

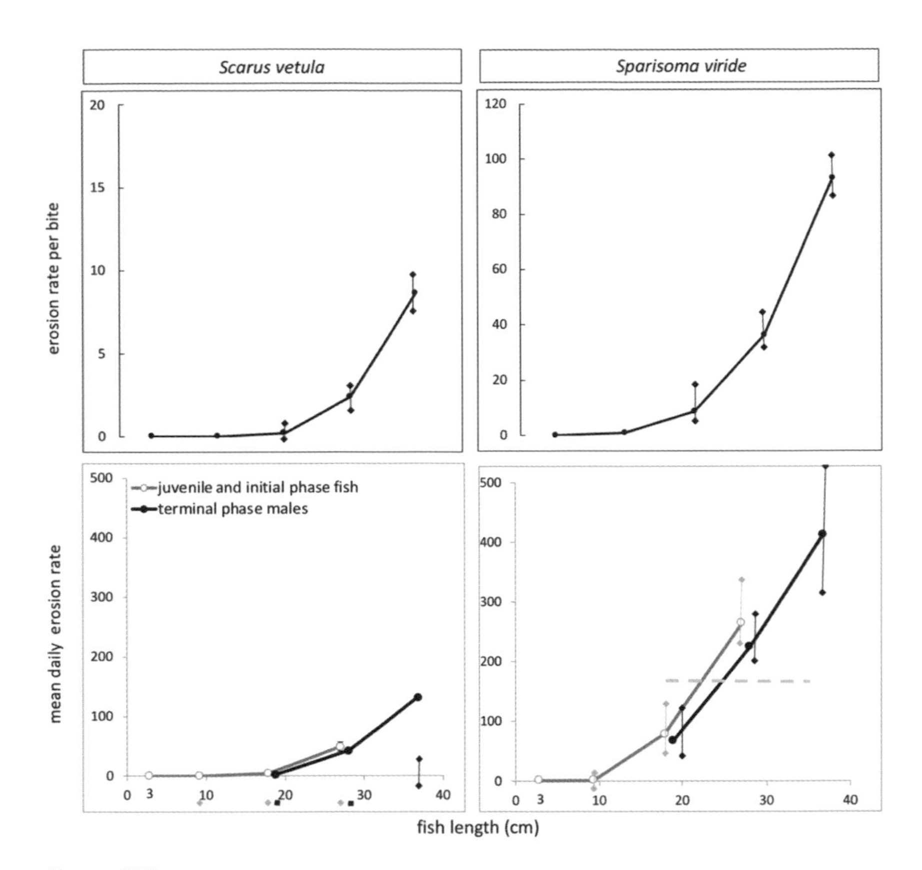

Figure 17.8

Comparison of the parrotfish *Scarus vetula* and *Sparisoma viride* erosion rates per bite (upper graphs) and per day (lower graphs) by fish size. Note the differences in the scales of the y-axis between the two upper graphs. *Viride* fish take much bigger bites than *vetula*! The bites are measured in milligrams per bite.

Credit: Redrawn after figure 5 from Henrich Bruggemann, A. M. Van Kessel, and A. M. Breeman, "Bioerosion and Sediment Ingestion by the Caribbean Parrotfish *Scarus vetula* and *Sparisoma viride*: Implications of Fish Size, Feeding Mode and Habitat Use," *Marine Ecology Progress Series* 132, no. 1–3 (1996): 59–71; yellow highlighting added

[1 acre in square meters]

There are 4046.86 square meters in an acre. If we divide 4046.86 by 300, we should get an estimate of the number of stoplight parrotfish per acre (according to Choat). This number is 13.5—a little higher than we estimated before, but not a ridiculous number.

Research Lessons

1. *Sometimes research takes time.* I admit it: this one took me a long time. I'm guessing I spent about ten hours researching this, and I learned a huge amount along the way. (Not just about the sex life of parrotfish, but about why it's so difficult to estimate the total number of fish on the reef.)

But I figured out that each stoplight parrotfish generates around 0.6 pounds of sand per day (on average, middle depth of the reef).

2. *Using the scholarly literature is easier than you think; use Google Scholar and Google Books to search it.* To estimate all that data about parrotfish, I went back to the original papers (using Google Scholar and Google Books) to dig out the charts and original data.

3. *Your local library is often a great friend.* In some cases, it's kind of a pain to get to the original source document. As you saw, I had to use my local library to get to the original paper by Thurman. But once I had it, the rest of the story unfolded easily.

4. *Don't let the big words scare you off.* Reading this scholarly literature was sometimes slow going. I had to look up a lot of words. *Acanthurids, scarids,* and *fork length* were all new terms for me. You can't let this kind of thing bog you down. You can always ask Google for definitions. (Just type [define fork length] into Google to discover that this is "the length of a fish measured from the most anterior part of the head to the deepest point of the notch in the tail fin." That's a handy trick to know when you're wading through dense text.)

Overall, though, this is a nice example of what you might think of as pure sensemaking—that is, pulling together information from multiple resources and cross-checking them against each other, and then checking facts against each other until the whole thing "makes sense."

In truth, I read much more than I've reported on here. I didn't want to tell you that I actually read about ten papers, some of which were duds. But this is also deeply true of real research. The papers seemed good from the abstract, yet the reality was that they encoded the data in an unhelpful way or were about parrotfish on some other continent or the wrong kind of fish altogether. When you do research from scientific papers, you have to look at the full paper, and not just the abstract. You might see just what you want to see in the abstract, and miss something deeply important that's mentioned in the details.

One of the traits of a great searcher is the ability to stick with it and persist in the search until you reach a satisfying result. And it *is* satisfying once you figure it all out!

More Reading for the Budding Ichthyologist

For comparison purposes of parrotfish feeding and sediment production on the Great Barrier Reef, see some of these papers. They're pretty technical, but if you're going to figure out something like this, it's inevitable.

Bellwood, D. R. 1995a. "Carbonate Transport and within-Reef Patterns of Bioerosion and Sediment Release in Parrotfishes (Family Scaridae) on the Great Barrier Reef." *Marine Ecology Progress Series* Marine 117:127–136.

Bellwood, D. R. 1995b. "Direct Estimate of Bioerosion by Two Parrotfish Species, *Chlorurus gibbus* and *C. sordidus*, on the Great Barrier Reef, Australia." *Marine Biology* 121:419–429.

Bellwood, D. R. 1996. "Production and Reworking of Sediment by Parrotfishes (Family *Scaridae*) on the Great Barrier Reef, Australia." *Marine Biology* 125 (4): 795–800.

Bruggerman, J. H., A. M. van Kessel, J. M. van Rooij, and A. M. Breeman. 1996. "Bioerosion and Sediment Ingestion by the Caribbean Parrotfish *Scarus vetula* and *Sparisoma viridae*: Implications of Fish Size, Feeding Mode, and Habitat Use." *Marine Ecology Progress Series* 134:59–71.

Choat, J. H., et al. 2003. An Age-Based Demographic Analysis of the Caribbean Stoplight Parrotfish *Sparisoma viride*." *Marine Ecology Progress Series* 246:265–277.

Gygi, R. A. 1975. "*Parisoma viridae* (Bonnaterre), the Stoplight Parrotfish, a Major Sediment Producer on Coral Reefs of Bermuda." *Eclogae Geologicae Helvetiae* 68: 327–359.

Thurman, H. V, and H. H. Webber. 1984. "Benthos on the Continental Shelf." In *Marine Biology*. Columbus, OH: Charles E. Merrill Publishing.

van Rooij, J. M., et al. 1996. "Resource and Habitat Sharing by the Stoplight Parrotfish, *Sparisoma viride*, a Caribbean Reef Herbivore." *Environmental Biology of Fishes* 47 (1): 81–91.

Try This Yourself

This chapter has shown you how to explore some unexpected animal behaviors in the natural world by following clues in your online research.

One of the more surprising things I've seen while out hiking are gigantic swarms of ladybugs.

Can you find out why and where these clusters of ladybugs take place? How big can these clusters of ladybugs get?

Figure 17.9
Ladybugs sometimes cluster together in immense groups that cover rocks, trees, and benches—any place where they can be together over the winter.
Credit: Amit Patel, by attribution, https://www.flickr.com/photos/amitp/

Of course, one of the potential problems with grouping together in such vast quantities is that dire things can sometimes happen as well. Do swarms of ladybugs ever end up in the water or on the shores of lakes? How would that happen?

Start with what you've read here to do your research. I suspect you'll quickly learn that there is more to the ladybug's life that simply eating aphids from your rosebushes!

18 Did Perry Ever Visit the Island of Delos? How to Follow a Long Chain of References to the Ultimate Answer

A surprising nineteenth-century graffiti leads to a world of discovery. How to find original, archival documents, and make connections over time, space, and history.

The Greek island of Delos is a beautiful, remarkable place that's full of classical history. It's a stony, sun-washed, smallish island in the Cyclades chain, more or less dead center in the circle of the Aegean Sea, a hundred miles southeast of Athens as the seagull flies and about the same distance from the Turkish coast.

It is also one of the most important mythological, historical, and archaeological sites in Greece, with archaeological excavations that are gigantic, as befits the birthplace of Apollo and Artemis. It was also the center of the Delian League and Delian Festivals, held every four years in a pan-Greek athletic competition that alternated with the somewhat more famous Olympic games.

When I visited the island, it was on a perfect fall day—with an endless blue sky, and deep-blue Aegean Sea. It was a day where you could easily imagine that the island was full of magic at the perfect intersection of land, water, and sky.

As the ferry approaches the short pier, you can see LOTS of temples and ancient buildings covering the hills as you come near the island. It is, you could say, thickly marbled with classical Greek ruins.

It's really quite a place. As you can see in figure 18.2, nearly the entire island is one giant archaeological site. People have been coming this way for thousands of years, both for religious and commercial purposes. They've

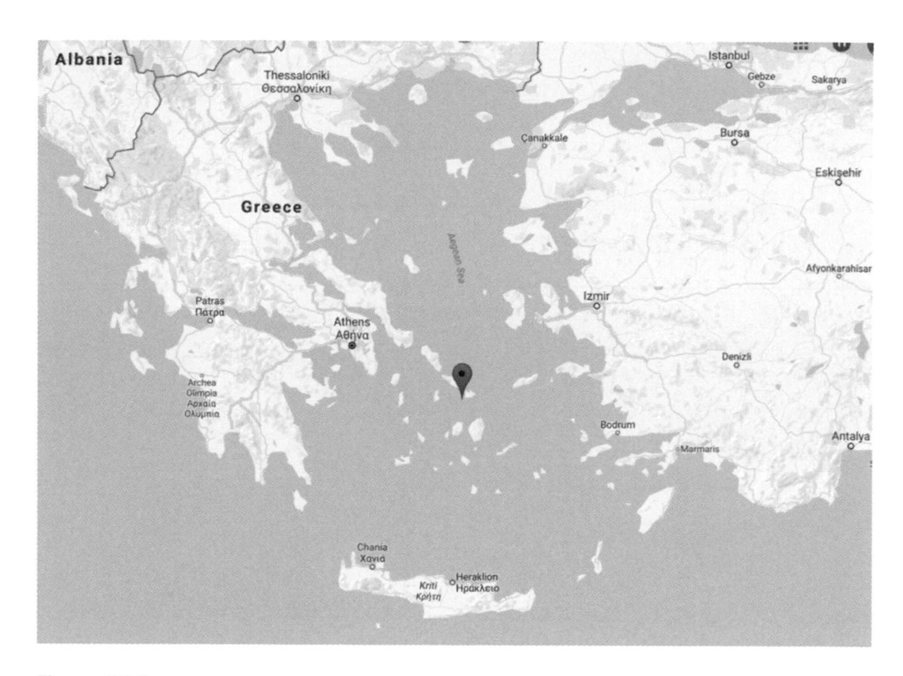

Figure 18.1
The island of Delos is in the center of the Aegean, the crossroads for every empire that inhabited the Mediterranean.
Credit: Map data © Google

been building homes, offices, markets, theaters, and temples all that time.

As a consequence, Delos was a stopping off point for traders, priests, slavers, worshipers, and travelers of every kind, pausing there to sell, buy, barter, steal, or worship, as was each traveler's particular bent.

But the island was attacked in 88 BCE by the Persian troops of Mithridates VI, an enemy of Rome, who killed nearly all the twenty thousand Romans on the island. Another attack came from pirates nineteen years later in 69 BCE, essentially wiping out the remainder, which knocked Delos out of action as an active trading island. By the end of the first century BCE, trade routes had changed to bypass Delos. Around this time the island became uninhabited and left as a place that the curious (and the occasional band of pirates) would visit.

When I visited in late summer 2016 (a couple thousand years after all the classical excitement), I toured the island with a guide, marveling at the ancient buildings, theaters, and temples.

Figure 18.2
Approaching Delos by ferry in 2016.
Credit: Daniel M. Russell

The tour guide did what tour guides do: walking us all around—and then pointed to this graffitied statue base (figure 18.4), saying that it used to hold a tall statue of Apollo, which was carried away by pirates sometime in the sixteenth century, leaving behind only the base. This was remarkable. I believe that there was a giant statue there, but I also know that we don't have any idea what it looked like.

But when I looked at the graffiti, I did a big double take. There was something so obvious and remarkable on that statue base that I couldn't believe the guide didn't say anything about this. Does it capture your eye as much as it did mine? In the center you can see that someone pecked out "Cap. M. C. Perry USN 1826" in the stone.

What's remarkable about this graffiti by M. C. Perry? Is it really *that* M. C. Perry? What are the chances?

This is clearly graffiti from a while ago (fresh graffiti on marble looks much whiter and cleaner). If you look at the stone carefully, you can make out a number of different names and dates. A few that stand out are:

B. Cooper, Esq., USN, 1826

John A. Cook, USN

Cap. M. C. Perry, USN, 1826

That last name rang a bell for me. As I stood there in the brilliant Greek sunshine, I wondered if I remembered my global history correctly: Wasn't

Figure 18.3
The central courtyard of a wealthy businessman. Note the circular cistern access in the near corner of the courtyard. Beneath the courtyard mosaic is a large, rectangular cistern for collecting rainwater.
Credit: Daniel M. Russell

he the commodore that forced Japan to reopen itself to trade in the 1850s after more than two hundred years of self-imposed isolation?

This practically dropped a curious question into my lap.

Research Question 1: *Is it possible that our tour guide missed the most famous graffitied name on the island of Delos? Could this possibly be the same guy as the naval commodore? Could this graffiti be from that famous M. C. Perry?*

Naturally, I started off by doing searches for all these graffitied names (Cooper, Cook, and Perry), each one at a time, but only M. C. Perry got many hits. It turned out that my memory served me well. M. C. Perry (full

Figure 18.4
Although our tour guide didn't point it out, there is a historically interesting bit of graffiti here. M. C. Perry pecked his name out on this plinth in 1826.
Credit: Daniel M. Russell

name: Matthew Calbraith Perry) was indeed an officer in the US Navy from 1809 to 1858.

The Wikipedia article about Perry tells us that *"from 1826 to 1827 Perry acted as fleet captain for Commodore Rodgers. Perry returned to Charleston, South Carolina for shore duty in 1828."* From this we learn that he was a captain in 1826, but otherwise that's a bit vague. It's not clear where he was in 1826–1827. Where was the fleet in those years? Where was Perry?

My first search to answer this was for:

["Matthew C Perry" Greece 1826]

If that hadn't worked, I would have started varying his name, just as we learned to do in chapter 16, trying variants like "M. C. Perry" or "Matthew * Perry"—but I didn't need to, as this query worked just fine.

The first hit was to an article from the official US Navy history archives that includes a biography of Perry and fairly extensive timeline. One entry in that relatively massive collection is this description of his service:

> Sep. 1824—5 Aug. 1827—Served as Executive Officer, U.S.S. North Carolina, flagship of the Mediterranean Squadron under Commodore Rodgers, engaged in protesting American commerce from Greek pirates. In 1825–26 participated in a visit to the headquarters of the Greek Revolutionists and in an interview with the Captain Pasha of the Turkish Fleet. Promoted to Master Commandant, 21 Mar. 1826.[1]

That's pretty compelling. We now know that he was in the Aegean Sea during 1826. Looking a bit deeper in the SERP, I found the Robinson library article about Perry, where it is said that "he subsequently [after 1824] served as Executive Officer of the USS North Carolina, flagship of the Mediterranean Squadron under Commodore Rodgers, which was engaged in protecting American commerce from Greek pirates. He [Perry] was promoted to Commander on March 21, 1826, and spent most of the next four years stateside."[2]

Figure 18.5 is confirmation of the dates, the ship, his officer, and his promotion to commander. I've made real headway on the research question. Now let's get a bit more detail here: Did the USS *North Carolina* (and Perry) make it to Delos in 1826?

To answer that question, we need to find some kind of book on the history of the USS *North Carolina*, or perhaps letters from men on board that tell us where they were on these dates in 1826.

I already had a big hint that this kind of thing existed when in the earlier searches, I read about the USS *North Carolina* in the Naval History and Heritage Center's History of the USS *North Carolina*, and saw the following image.[3]

Figure 18.5 is the frontispiece of the USS *North Carolina* logbook from US Navy military history site. This is the log of the right ship, but it covers the wrong years: 1836–1837. In the case of Perry in the Aegean, I'm looking for 1826. Not being an expert in nineteenth-century sailing procedures, I hadn't thought about searching for a logbook! But finding this logbook of the USS *North Carolina* tells me that there's a chance we could find the logbook of the ship for 1826 in a library somewhere. Now can I find the logbook of the ship for 1826, and will it show that Perry was on Delos on the date that's in the graffiti?

Figure 18.5
Finding the logbook of the USS *North Carolina*, even though it's for the wrong year, reminded me that navy ships have logbooks and that I should search for the logbook from 1826.
Credit: Photo Archives, Naval History and Heritage Command, NH 65196

My new research goal is to find the actual logbook from the USS *North Carolina* from 1826—and if I find it, can I actually read it?

Research Question 2: *Where is the logbook of the USS* North Carolina *for 1826?*

I have no idea where the logbook might be, so my query was broad:

["logbook" "USS North Carolina"]

And I found that the first result was from Google Books—the *Guide to Non-Federal Archives and Manuscripts in the United States Relating to Africa: Alabama–New Mexico*.[4]

That unwieldy title tells me that this book is a *finding guide* (sometimes also called a *finding aid*), which is an index to archives and manuscripts in other places. It's not the books themselves but instead a kind of catalog of places where you can find these books in other libraries. In this case, it's an index to the archives and manuscripts that are not owned by the federal government, and located in the states from A–N. This suggests there's another finding guide for states from NY–W … and that there's probably another couple of finding guides for *federal* archives and manuscripts. That's

Page 217

in Chief in the Mediterranean; a letterbook kept on U.S.S. North Carolina, July 1824-July 1827, containing fair copies of letters sent by John Rodgers as commander of the Mediterranean Squadron; a two-volume journal of a cruise in the Mediterranean of the U.S.S. 74-gun ship North Carolina, September 1825-August 1827; and the logbook of U.S.S. North Carolina, January 1826-April 1827.

Figure 18.6
Credit: H. Zell Publishers

a handy thing to remember for the next time that we're looking for some archives. Sometimes the feds really do have the documents, and it's worth remembering that the term *finding guide* is a useful thing to search for.

By looking in the Google Books copy of this finding guide, I ran across an intriguing entry on page 217 after doing the obvious search within the book for "USS North Carolina" and "logbook" (figure 18.6).

This is great! Not only do we now know that this logbook still exists somewhere; this finding guide can tell us where the logbook from 1826 is actually kept. Notice that this isn't just the logbook but this archive also has a *letterbook* (that is, "fair copies" of letters written by the fleet commander, John Rodgers).[5] With luck we'll find a letter that mentions their visit to Delos, or at very least, we'll find a logbook entry that talks about a stop at Delos.

Yet the digitized copy of the finding guide in Google Books is only in "partial view"—meaning that while the entire book is digitized, not all of it is available for viewing in the browser because of copyright restrictions. Rats. This means that while I know the logbook exists, I can't read far enough in the online text to find out where! That's fine, since I'm willing to switch to doing off-line research, but this means I've got a new research question to work on.

Research Question 3: *Where is this finding guide?*

I need to physically locate the *finding guide* so I can read the rest of page 217 and discover where the logbook is actually kept.

Luckily, that's easy. I clicked on the "Find in a Library" button (on the left side of the Google Books page; see figure 18.7) to show me where copies of the finding guide are kept. Google Books has a deal with WorldCat (the worldwide aggregated catalogs of many thousands of libraries) that will let us locate this finding guide book wherever copies of it are kept.

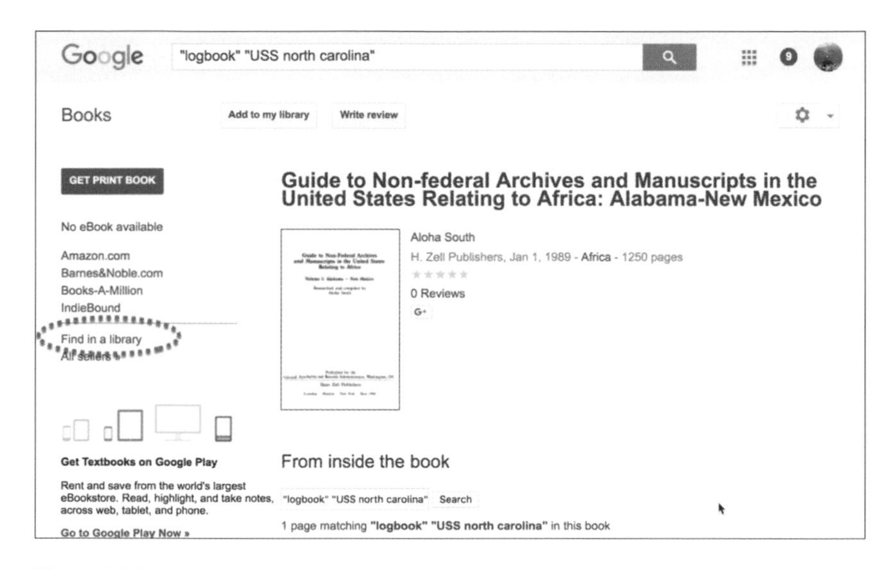

Figure 18.7
Circled in the dotted red oval is the link to help find this book in a local library. Clicking here takes you to the online library metacatalog that has the physical locations of thousands of books worldwide. By entering your location, it can tell you the closest physical copy to your location.
Credit: Google and the Google logo are registered trademarks of Google Inc., used with permission

After clicking on the "Find in a Library" link and entering my zip code to locate the nearest copy, to my delight I discovered that this finding guide is kept at a library just two miles from my house at Stanford University! It's an amazing coincidence that the finding aid would be just a bike ride away. If I can get to the finding guide at Stanford, I'll be able to look at the rest of the page to figure out where the logbook actually physically resides.

The next day I called up the Stanford library and asked if I could read page 217 of the *Guide to Non-Federal Archives to …* They were kind and said, "Sure, can you come tomorrow?" (And this, I quickly learned, is common practice among archives. Many are open to the interested public. You often don't need a special membership.)

At the right time, I bicycled over to the library, went to the rare books room, and checked out the finding guide, turning carefully to page 217 to find that the logbook for 1826–1827 is kept in the Elutherian Library in Wilmington, Delaware. I did a quick search to see if any of its collection is digitized, but no luck; the Elutherian doesn't seem to have much of its collection scanned and available online. On the other hand, I did discover that the Elutherian Library merged with the Hagley Museum and Library in 1984. I'm probably not going to ride my bike there, but on my next business trip to the East Coast, I might be able to swing a trip to visit the Hagley library to see the logbook and perhaps the letters file as well. Perhaps that will shed some light on Perry's presence (or absence) on Delos.

As I was thinking about this, it occurred to me that it might be possible to find a record of the movements of Perry's superior officer, Commodore Rodgers, of the Mediterranean fleet in 1826. (This relationship between Rodgers and Perry was a fact that I picked up while reading the Wikipedia entry about Perry. I learned that his commanding officer was Rodgers, who was quite a historical figure himself.)

To find out if Rodgers had ever visited Delos, I did this query in Google Books to search for a book with information about their visit:

[Navy John Rodgers Delos]

I was rewarded with a hit! In the book *The Navy's Godfather: John Rodgers* by Eileen Lebow, I find this quote at the beginning of chapter 16.

> By the middle of June, 1826, the squadron had reached Vourla near Smyrna, after a leisurely thirty-four day sail, the length of the sea, from Gibraltar with stops at Algiers, Tunis, Carthage and the islands of Milos, Paros, and Delos. The commodore spent four days at each island, digging among the broken columns and tombs, exploring what were once magnificent Greek structures, and collecting enough relics to fill ten wagons, including two altars from the temples of Diana and Apollo on **Delos**.[6]

Not only do we know that Commodore Rodgers and the USS *North Carolina*, which had Perry on board, visited Delos, but they specifically stopped at the temples of Diana and Apollo. (Recall that the graffiti was found on the base of the statue of Apollo … at his temple.) I'd say that Perry, the famous one, was not only there but also spent quite a bit of time etching his name into the pediment at the Temple of Apollo.

How can we find out more about Perry himself? I just tried the obvious (although rather broad) search for items about the US Navy in 1826:

[U S Navy 1826]

In the results (on the first page!) I found the *Naval Register for 1826* (the register is the master list of all men in the US Navy), which includes entries

Figure 18.8
A painting of M. C. Perry from about 1853. A photograph of a painting by Orlando S. Lagman, US Navy, based on an earlier painting done about 1853.
Credit: Photo Archives, Naval History and Heritage Command, NH 47490

for Rodgers, Cooper, Cook, and Perry.[7] This is another confirmation that they were all on the same Mediterranean cruise serving under Rodgers and apparently "visited" the Temple of Apollo together as well. They were: John A. Cook, lieutenant on the *Porpoise*, which was sailing alongside the *North Carolina*; Thomas J. Manning, a midshipman on the *Porpoise*; Benjamin Cooper, a lieutenant on leave; and of course, M. C. Perry, who was (at the time) a master commandant on the *North Carolina*.

All I had to do now was make it to Delaware to visit the archives.

A few months later, on a chilly, gray, rainy day, I stepped off the train in Wilmington and hopped into a taxi to the Hagley library. It's a few miles away from the train station, and the ride there was to a tony section of

Wilmington—with large manicured lawns, leafy trees, gracious mansions, and lovely stone buildings set next to Brandywine Creek.

The archivists were kind, and once I'd given them the accession number from the finding aid, they set me up in a corner of their quiet reading room with two boxes of archival content from the USS *North Carolina*'s travels during 1826 and 1827. These boxes held the logbooks and letters.

I spent a couple hours reading through the logbook, and learning via the letters about what life was like on board a ship of the line in the Mediterranean in 1826. There are inventory lists, judicial actions, updates in letters from the United States, and endless notes about weather conditions and ship movements.[8]

As interesting as that was, my goal was to find out if the USS *North Carolina* actually visited Delos, and if so, on what days.

It took a while, but after reading for a bit, I was happy to find this page in the logbook, dated June 15, 1826, when the ship was "standing towards the south side of Delos" for several hours (figure 18.9).

Interestingly, I wasn't able to find any other visits to the island in the logbook, let alone a visit of four days to Delos, Paros, or Milos. But it's clear from this record that the USS *North Carolina* lingered for several hours on Delos, and the ship's crew visited other nearby islands as well. There's no comment about whether they were able to off-load "enough relics to fill ten wagons." On the other hand, they were certainly there long enough for Cooper, Cook, and Perry to peck their names and dates into a marble plinth for me to find nearly two hundred years later.

But I didn't yet have explicit confirmation that they went onto the island itself. How can I find something that makes it clear they went onshore?

Every researcher has a moment when the course of your investigation changes. In this case, one of the librarians at the Hagley said, "You know, in those days, many ships had more than one logbook. You might check around for another copy."

It had never occurred to me that there might be *another* copy! As I said, I'm not an expert in nineteenth-century navy sailing ships or the peculiarities of the way that they kept their logs.

So going back online, I repeated my search from above:

["logbook" "USS North Carolina"]

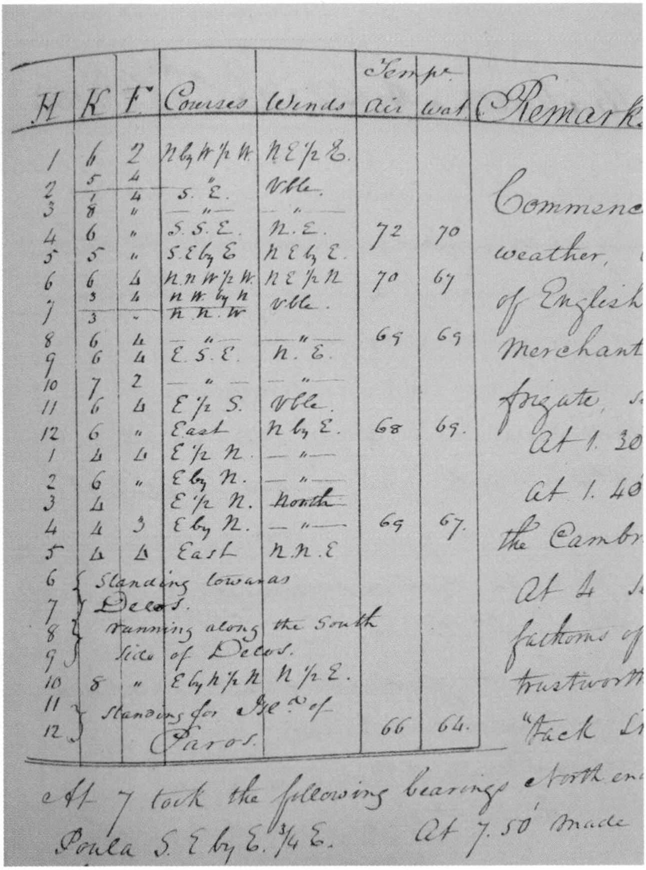

H	K	F	Courses	Winds	Temp. Air	Temp. Wat.
1	6	2	N by W ½ W	N E ½ E		
2	5	4		Vble		
	1	4	S. E.			
3	8	"	— " —	— " —		
4	6	"	S. S. E.	N. E.	72	70
5	5	"	S. E by E	N E by E		
6	6	4	N. N W ½ W.	N E ½ N	70	67
	3	4	N W by N	Vble		
7	3	-	N N. W			
8	6	4	— " —	— " —	69	69
9	6	4	E. S. E.	N. " E.		
10	7	2	— " —	— " —		
11	6	4	E ½ S.	Vble		
12	6	"	East	N by E.	68	69.
1	4	4	E ½ N.	— " —		
2	6	"	E by N.	— " —		
3	4		E ½ N.	North		
4	4	3	E by N.	— " —	69	67.
5	4	4	East	N. N. E		
6–7	Standing towards Delos.					
8–9	running along the South Side of Delos.					
10	8	"	E by N ½ N	N ½ E.		
11–12	Standing for Isle of Paros.				66	64.

Remarks
Commenc…
weather, …
of English
Merchant
frigate, …
At 1. 30
At 1. 40
the Cambr…
At 4 …
fathoms of
trustworth…
"Jack …

At 7 took the following bearings North en…
Poula S. E by E. ¾ E. At 7. 50' made

Figure 18.9

As the logbook from June 15, 1826, notes, the USS *North Carolina* was alongside Delos for much of the late afternoon.

Credit: Papers of the USS *North Carolina*, accession 383, courtesy of the Hagley Museum and Library

When I read further down in the results, I discovered at least two other libraries that have "logbooks" for the USS *North Carolina*. Really? I'd always thought that there was just one, so when I found the first result, I stopped, assuming that I'd discovered the logbook. But no, it turns out that there were often multiple logbooks, frequently written and maintained by different people on board.

The second logbook is in the archives of the University of North Carolina at Chapel Hill. It's listed in that archive's finding aid as accession number 03190-z (with the collection title North Carolina Logs, 1825–1827).[9] Once again, it's in the library, but also not digitized!

Luckily, I have a friend in Chapel Hill who was happy to run over to the library and investigate its copy of the logbook.[10] Sure enough, its copy of the log contains the same information, although clearly written by someone else; it's not an exact copy of the Hagley library logbook. There's also no direct confirmation that sailors from the USS *North Carolina* actually went ashore.

Encouraged by this, I went on to find that the third logbook is at the Library of Congress in its archives, filed in the Rodgers Family Papers collection, 1740–1987.[11] In retrospect this seems obvious enough, but at the time, this was a perfect example of a serendipitous discovery.

As luck would have it, I was planning a trip to Washington, DC, just a few days later. I contacted the library's archive section and asked if it would be possible for me to see the collection. Once again, the archivists were gracious, and on another rainy day in May, I visited the great marble edifice of the library's Madison Building, just down the street from the US Capitol.

Once I got my reader's card, I asked for a box that contained the logbook and told the librarian at the desk about my quest. He nodded and asked me to fill in a requisition slip (in pencil because they don't allow pens in the archive room). I did that, asking just for the archive box with the logbook. He looked at my slip and compared it to the online finding aid, commenting that "maybe you'd like to see the letters of John Rodgers as well. … I suspect you'll find them interesting."

I learned that when an archivist says something like that, you really should listen, so we ordered another three boxes of the collected letters and notes from the collection.

When the four boxes arrived at my desk, I realized that while this collection has been used before, so far as I could tell, none of it has ever been photographed or scanned for online research.

Naturally, I turned to the logbook to June 15, 1826, and found more or less the same information as I'd seen in the other logbooks. I now know that the language of the logbook is fairly formulaic; standard phrases abound, and I didn't learn anything new from the third logbook.

On the other hand, the archivist's suggestion let me look at these three boxes of missives from 190 years ago. Taking the archivist's cue, I started reading the letters of Rodgers to his wife, Minerva, and her replies. In many ways, it's like reading an email thread; there were both halves of the conversation—his letter, and then her response, back and forth over the years while he was sailing around the world.

Mostly this exchange is about the little things about life on board the ship or life with the children back in Washington, DC, where Minerva lived.

Finally, I found a letter from Rodgers that's dated June 26, 1826. This is what I discovered on the bottom part of page 1 (figure 18.10).

> Of all the professions I suppose you would think me the least qualified for that of an antiquarian, notwithstanding this I have at the three last named places employed from fifty to sixty men daily with pick axes and spades in search of antiques & if I have not found as many relicks I have at least opened as many tombs & exposed columns, capitols & friezes & for that lay hidden in the bowels of the earth as most of the antiquarians of the age—I have not as yet been so fortunate as to find anything that can be said to be universally very remarkable, although some no doubt that will be considered rare in our country—I have at this time onboard the NC [the USS *North Carolina*] as many relicks of one kind or another as would load ten waggons & among them number two white marble altars from the Temples of Apollo & Diana at Delos.

That seems pretty compelling evidence that Rodgers and Perry (his close friend and fellow officer) spent enough time at Delos to do a bit of work, plundering the island of its classical treasures and leaving behind a bit of graffiti marking their time on the island.

Research Lessons

There are a few key lessons here.

1. *Sometimes you have to go off-line to complete your research.* This is a book about online research, but the reality is that not everything is digitized and available online. There are enough clues in the online world, however, so I was able to find the physical logbook and then look at the actual record of events, written in elegant longhand on board the USS *North Carolina*.

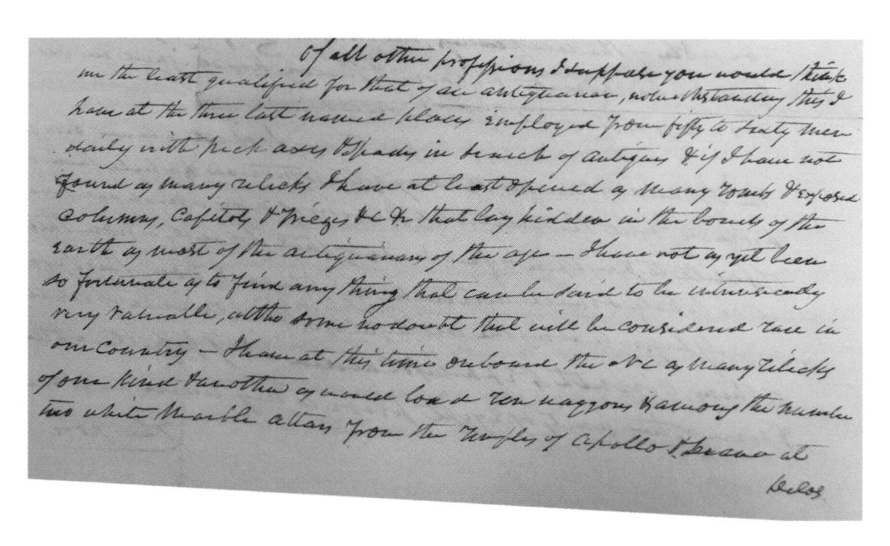

of all other professions I suppose you would think
me the least qualified for that of an antiquarian, notwithstanding this I
have at the three last named places employed from fifty to sixty men
daily with pick axes & spades in search of antiques & if I have not
found as many relicks I have at least opened as many tombs & exposed
columns, Capitols & friezes &c &c that lay hidden in the bowels of the
earth as most of the antiquarians of the age — I have not as yet been
so fortunate as to find any thing that can be said to be intrinsically
very valuable, altho some no doubt that will be considered rare in
our Country — I have at this time onboard the N C as many relicks
of one kind & another as would load ten waggons & among the number
two white marble altars from the Temples of Apollo & Diana at
Delos

Figure 18.10
In his letter home to his wife, Minvera, John Rodgers writes about collecting antiquities, such as "*ten waggons & among them number two white marble altars from the Temples of Apollo & Diana at Delos*" (June 26, 1826).
Credit: Library of Congress

Yes, more and more stuff IS being digitized, but many archives don't have the resources to move everything into the digital realm. So sometimes you have to visit the archives in real life. That's how I ended up in Wilmington on a rainy afternoon visiting the original copy of the USS *North Carolina* logbook.

2. *Follow the chain of references*. In essence, this chapter is a story about going from a query, to articles about Delos, to discovering that logbooks exist, to using a finding aid to locate the logbook, to physical archives. Once I found the finding aid that pointed to the logbook, I had to go to the Stanford rare book room to find the accession number for the logbook in Delaware, then to the Hagley library in Delaware to read the logbook, and then realized that I needed to find better evidence. I found other logbooks, and finally discovered the finding aid for the Rodgers Family Papers collection, going in person to the Library of Congress to finally find a personal letter confirming that they were on Delos. That's quite a trail, with a mix of online searching and off-line reading.

3. *If you keep digging, you can find remarkable things.* I didn't know that we'd be able to locate the logbook of the ship that took the graffiti artists to Delos. This is a nice example where doing one more query takes your research to a whole new level. Even better, I didn't realize that the trail would lead me to reading nineteenth-century personal letters in order to uncover what really happened on Delos.

4. *Remember to follow parallel paths to get to what you seek.* In the above example, the trick that got us to the logbook of the USS *North Carolina* was to do a search for Commodore Rodgers, who we KNEW had to be on board with Perry. That kind of parallel searching is incredibly useful (when you can do it). Sometimes the shortest path to your goal is to find someone (or something) that must also be closely related to what you're searching.

5. *Serendipity favors the prepared mind.* This research voyage into the tour of Perry began with my spotting his name and date on the stone. Before going to Greece, I knew something about Perry; his name was well known to me, not because of Greek history, but because of what I know about Japanese history. His name is well known in world history, yet it wasn't a straight path from Delos to the archive in Wilmington—although I'm glad that I followed the path.

6. *Learn as you go.* As you read, notice things that might be useful later on in your searching. When I was studying Perry, I learned about Rodgers. That chance "learning while searching" led me to the Library of Congress archives and that letter from Rodgers to his wife telling the story of being on the island, confirming that Perry was on Delos.

7. *Go a bit deeper than you think you need to go.* I was surprised to learn that the first logbook was great, but that both the second and third logbooks were even better! When you're doing research, it's often worth searching to the bottom of the page of results, and for historical research like this, maybe even going one page deeper into the results. You might find buried treasure!

Try This Yourself

Almost every city and town has an archive somewhere nearby, and they're often full of images and stories from your town, your city, and the lives of the people around you.

In the course of this chapter, I ended up using four different archives (Stanford's rare book room, the Hagley Museum and Library, the archives at the University of North Carolina at Chapel Hill, and the Library of Congress archives). All I did was ask nicely if I could use their archival content. Sometimes you have to get a reader's card (just like a library card), but in every archive I've visited, I've never been turned down.

So explore your local archive; do a quick Google search for [archive *city name*] for the archives near you. Once you find one, check first to see if they've got any online content that you can easily access. (Checking their online holdings can frequently save you a trip to a closed place or can reveal to you the treasures that they have ready for you to investigate.) Once you've found one, they can often direct you to the next, and the next.

Some sample questions to ask of your hometown: Can you find the early history of your town? In my city, the main library has a beautiful archive of early city historical imagery and maps. Once you can see into the past, research questions will spring up in your mind easily. When and why was this city founded? Why did the founders choose *this* particular place? How was the original grid of city streets established? Or was it all organic, following old paths and trails lost in the mists of time?

19 On Being a Great Searcher: Rules of Thumb for Asking Great Questions

What does it take to be a great online searcher? Here's a set of observations about what I've seen great researchers do. First, you have to understand what it is that you're asking, both in terms of what the question is about (in your opinion) and what you're learning as you do your research. But there's more—much more—to being really good at searching.

I need to point out a few things I've learned that successful researchers all share.

I've summarized these observations about how to do great research into three sets of observations: behaviors (what they do when doing research), attitudes (the way that they deal with setbacks and the search process), and rules of thumb for asking good questions.

As I've watched the way that truly expert searchers work, I've noticed how they frame their investigations, what attitudes they have in common, the manner in which they follow up on ideas, and the way they get to answers.[1] (And to be honest, I've also watched the way that a lot of people make serious mistakes as they search. Those negative lessons are valuable too, if only to illustrate what NOT to do.)

These aren't just three random lists of good practices to do while you research. They're based on the past thirteen years of my own studies, where I've watched thousands of researchers trying to answer questions. I've watched professional fact-checkers, journalists, and just plain folks trying to search for good answers to their questions. Whether you're investigating a breaking news story or trying to determine the best refrigerator to buy, we all do much of our research online—and these points summarize the way that you can be fast, accurate, and reliable in your research. Here are the behaviors, attitudes, and rules of thumb for being a great researcher.

Behaviors of Great Researchers

Here are a few things that are common among all the really good searchers. This isn't a complete list, but they reflect the patterns that I've seen most often.

1. *They figure out what the question really is.* As you've seen in each of the chapters so far, I start with a bit of a curious story and then translate that into a research question. An important skill is the ability to see a problem and convert it into a researchable question. That's why I've been giving the research questions at the beginning of each chapter. Learning how to convert a vague feeling, hunch, or lingering sense of not understanding into a research question is an important step.

Even if you don't write down your research question, it's useful to pause for a moment to clarify what you seek. A common problem among beginning searchers is that they're not clear about what the goal is and start searching a bit blindly.

A good research question is also something that you, the researcher, really understand. A great mental checkpoint is to ask yourself, "Do I really understand what my research question means?"

Often, you'll have to learn to DEFINE terms in your questions. Take the research question, "How far is it to the Moon?" You'll want to clarify that question a bit. Do you mean, "The center of the Moon to the center of the Earth?" or "From the closest point on the Moon to the closest point on the Earth?"

Or suppose your research question is "I want to learn about the Civil War." That's great, but fairly open ended. WHICH civil war? There are a bunch; luckily, search engines figure out where you are and give you the most common interpretation of civil war. But in some countries, that's not obvious; perhaps you meant the most *recent* civil war? If you're in the United States, does your question mean the US Civil War or the English Civil War? (And if the latter, *which one?* Inconveniently, England has had multiple wars of incivility.)

This is an important point. You must be clear about what you're asking—otherwise you'll find an answer to a question that might not mean what you think.

In chapter 10, we had to figure out what "first oil discovered" really meant. Was it the "first to notice an oil-like substance"? Or do you mean the "first to pump it out of the ground"? Or the "first to produce oil products (like gasoline) for sale"? In a real sense, clarifying the question is the primary research strategy.

2. *Great researchers recognize what they're doing as they search; they understand the problems, questions, and areas of uncertainty and doubt.* One of the things that struck me when watching great researchers is that they think about their research questions fairly explicitly and clearly as they do their searching. If you interrupt one of these folks, they can tell you what they're doing at the moment and why. (In the chapter on searching for Perry's graffiti, I could give you a good reason why I went this far out on a research limb. Why did I go to the archives at the Library of Congress? Because "I'm searching for the Rodgers Family Papers at the library archives because I need to find out what the crew did while on Delos, and I suspect there's a document there.") But by the nature of research, you don't know how well it's going to turn out. Great researchers also follow plenty of clues that turn out to be dead ends. Nevertheless, they're usually pretty clear about what the research question is, and how much they know (at any point) about the target of their work.

This means that you, as a researcher, have to understand what you know, what you don't know, and how confident you are that what you DO know is accurate. What you do from there in your searching is largely determined by this balance of known and what you need to figure out next. Online research is often a process, meaning that you're moving from a place where you don't know much to one where you have good understanding. Your next query is largely determined by what you need to know next.

3. *They understand that they need to contextualize what they find.* One of the most common strengths of good researchers is that they don't just answer the question as stated but instead learn how to understand the context around the question.

What's context? It's the information that helps you understand what you're finding in a larger sense, usually by telling you what you *also* need to know to interpret the information that you're finding.

For instance, if you find that a restaurant has a score of 4.5 out of 5, that sounds great—but knowing that there are only three reviews is a bit of context that helps you understand that it might not be that impressive. If the review score is 4.5 over two thousand reviews, that's a different story.

Almost everything that you search for needs a bit of context to help you make sense of the information. In particular, when reading an article with statistical information, that almost guarantees that you'll need some context to understand the rest of the story. Instead of being shocked that the temperature was forty-five degrees yesterday, look for the longer trends and context that will help you interpret that. Was this temperature an isolated data point that's *really different* from other days? Perhaps most important, was that the temperature in Marrakesh (where that forty-five would be in Celsius) or San Francisco (where that would be a typical evening's temperature in Fahrenheit).

Great searchers do more than just find an answer to a research question; they also talk to subject experts, read papers and critiques in their area, and study articles that talk about the topic. They constantly ask themselves, "What else should I know in order to see how this information fits into a larger context?"

4. *They select* ***what*** *to ask questions about.* This is an important skill that's as much about good taste as it is about good research. You can't ask questions about everything, so how do you decide what to ask questions about?

Answer: basically, you ask questions about the unusual, the thing that's driving your curiosity. That is, you ask questions about the things that stick out for you—the odd, or the thing that strikes you as not quite right. This means that you need to know what's normal and what's not. For instance, in the oil well chapter, I say that there are more fruits and nuts grown in California than anywhere else. You might just accept that because it sounds fairly reasonable and comes in a form that seems believable (in particular, that it's in this book). But you would check it if you thought it was weird or at odds with what you already know.

For example, suppose I'd said something that you found really hard to believe—maybe, "There are more Armenians living in California than in Armenia." THAT's the kind of thing you'd want to check.

Of course, when you're in the middle of a search process, you're frequently on the trail, running down a piece of the story that you need to know about. In these situations, it's crucial to stay on track, and not get diverted by all the shiny and sparkling things that want to take you off the topic. That's yet another reason to keep your research question in front of you. Yes, there's always something odd and strange that might take you off the path, but resist that impulse; stay on your question until you complete the research. (If you must, take a note about things to return to later. I often do this while searching, and then when I look at the list later, I can't imagine why I would want to follow up.)

5. *They already know a bunch of stuff.* It seems paradoxical, but it's deeply true: the more you know about a topic, the more you can do research in it. That's because the more you recognize (terms, concepts, and relationships), the better you are at noticing what's important and what's not. Online research tools are truly excellent, but people who know more about a domain simply do better research in that domain.

For instance, in chapter 14 we learned about Key's family. Because I'd already read and knew about his son, Philip, I was surprised when I saw his name in a new story that didn't match the story I'd already read. That is, when you know something of the background (the people, places, events, and relationships), anything that sticks out (because you haven't seen it before) is probably interesting. It doesn't match the pattern you already know and so is possibly worth chasing.

The bottom line is that good researchers do well in part because they already know things about the world at large. In addition, they also understand what the tools can and cannot do.

6. *They understand coverage and limitations.* Researchers who use search engines and online research tools every day need to understand what's possible to do with them, and the assumptions that are built into them.

Search engine coverage is impressively large, but we have to overcome the illusion of omniscience (particularly with students learning to do their research online). It's worth remembering that what's searchable online is not the sum total of all human knowledge. Not everything is online and searchable. A great deal IS, but the further back you search in time, the more spotty the coverage grows.

While we, as a culture, are putting more and more content online, and more content is "born digital," there is still content that's off-line and will be unavailable for the foreseeable future. What's more, copyright and policy issues will keep content tied up in unsearchable ways, while corporate issues will continue to affect the availability of information.

In particular, great researchers recognize that there are differences between search engines and information resources. The large, general-purpose, broad-range search engines (e.g., Google, Bing, etc.) provide superb coverage and the ability to supply depth on topics of special interest. They're extremely good at covering web content, news, text resources, images, videos, maps, and so on. They are less good at providing in-depth search services in specialty topics (e.g., mathematics services, domain-specific providing context, etc.). The information landscape is not flat, nor are search engines completely universal in their coverage and competence. Each of the search engines offers a large number of different kinds of knowledge services to its users.

Due to local policy or legal restrictions on what kinds of knowledge can be served, search engines will always be slightly a bit different in their behavior from place to place. This isn't just an odd property of the way that they're built but rather a deep observation about the nature of social and political factors at work. Just as the content of encyclopedias was never consistent across national boundaries (for example, is Burr an American traitor or British hero?), so too will search tools necessarily serve different versions of knowledge depending on where the query is issued and what knowledge is received. Maps are currently different depending on where they're viewed (contested national boundaries always look different from the other side of the border dispute), and this is true for contentious data sets as well. Geoffrey Bowker and Susan Leigh Star illustrate this well in their book about how people organize knowledge about things: not only is medical knowledge organized differently from place to place, but its use and interpretations are as well.[2] Different countries and cultures really do see the world differently, which is why looking at different versions of Wikipedia (by language) is such an interesting way to see the same topic from a different perspective (see chapter 7).

7. *Understand how search interfaces work.* Currently, search engines are getting better at answering questions. But at the moment, there are still

questions that can't be answered automatically (e.g., [who was Daniel Russell doctoral adviser]). Even though you could figure it out by doing your own searches, one step at a time, search engines can't quite (yet!) put all the clues together to answer a question like that. (See the next chapter, though. Things are getting better all the time.)

The thing is, search engines don't signal that they lack the knowledge to supply an answer, yet they don't want to look bad, so they give a web-search set of results instead. That's a great fallback position, but it's also an important difference between an *answer* and a set of search results—one that's worth noticing.

A skilled searcher knows that the text abstracts (aka snippets) for each web result are algorithmically generated without deep semantic processing. Effectively, the snippet composition system selects out fragments of text that score highly with respect to the interpreted form of the query. Those fragments are then concatenated together with ellipses, sometimes leading to an unintended interpretation when read as a summary of the page. If you know this about snippets, the correct reading is clear and straightforward—but this model isn't explained, and isn't widely understood by beginning searchers.

In other words, a searcher has to learn to interpret the subtle signals that are often quietly expressed in the interface design. Mostly (through iterative design and testing many variations on a theme) the designers of the search engine have arrived at a solution that works for most people in most cases. But for critical readings and complex research tasks, the search engine interface needs to be read with some skill and understanding. This includes attending to changes in the user interface as well as stepping in and questioning search results when an error seems possible. Remember, just because it shows up in a search engine result page doesn't mean it's true; it just means that you found something.

When you're doing online research, you're using a particular set of tools, such as search engines, databases, and their ilk. Each of these tools makes some things easy and other things almost impossible. An incredibly useful behavior to have is the habit of *reading the user interface*—that is, exploring the set of abilities that the tool gives you. For instance, can you sort these results by date or only by location? Is it possible to search over all the data or only parts of it? Spending a few minutes when you first start using a tool is incredibly useful; it will tell you what kinds of things are possible and what ones are just super hard. Spending that time to learn is worth the investment because sooner or later, you well may need to use that ability to find something.

Skilled searchers ask good questions that match the capabilities of the search engine. Searchers need to be sophisticated about what they are asking and thus what kind of answer to expect; the world is complicated, and not all simple questions have simple answers. For example, when was the USS *Constitution* built? The keel was laid on November 1, 1794. The ship was first launched on September 20, 1797 (but it accidentally stopped short of the water). It finally landed in the water and was commissioned on October 21, 1797. Even simple questions can have unexpectedly complex answers.

The increasing sophistication in representing world knowledge online also implies that asking the right questions will become more of a skill. A common error made among beginning searchers is to pose queries that have a built-in bias, a kind of leading question. A question that "leads the witness" wouldn't be allowed in a courtroom and shouldn't be in your research queries either. This is fairly common among K–12 students who don't yet understand the basics of web search and often frame their questions with built-in assumptions. In cases like this, you need to know that the results are ordered by rank depending on the terms in the query. Take the following query:

[is the average length of an octopus 25 inches?]

It will give web links in the search results page that look right, but there's an assumption within the query—that is, that octopuses actually are, on average, 25 inches in length. The web search results will be a bit misleading because there are so many positive hits that mention the terms "octopus" and "25 inches" on the same page. In this case, the search engine doesn't *really* understand the question, but gives pages that best match the query, with its biases built in.

8. *They read boldly yet carefully*. Just as with the skill of reading snippets today, reading the answers generated by search engines carefully is an essential skill, especially when learning new user interface idioms that come into play. For example, for a simple question like [what are the languages of Eritrea], the answer will be displayed as "Eritrean Official Languages: Tigrigna, English, Arabic," even though six other languages (with large, distinct

populations) are also spoken there. If you miss the word "official" in the answer, you'll expect the answer to match your question and will miss the one million Tigre-speaking people of the country.

Careful reading includes more than just the text on the results; it also includes the little clues that you're given in the interface. For instance, if you do the query [how many states in India], you'll see that there are (as of 2018) twenty-nine states, although there are several sites that claim there are twenty-eight states. Reading carefully, you'll see that the latest state, Telangana, was formed in 2014 (websites that haven't been updated since then have it wrong). But there are also seven "union territories," which aren't quite states, but are regions that are ruled directly by the union government (central government).

Reading boldly just means that you do NOT let yourself be intimidated by scary-looking titles, big words, or content that seems too complicated. Using a couple of online research tools, you can read just about anything.

How can I read it and understand what's going on in complex texts? I have a two-step process for making this understandable.

First, I read through the article, searching for terms and concepts that I don't understand. I look up these terms, usually by opening new tabs with the searches, both so I won't lose my place in the original article and have several pages open for reference. My friend and colleague at Stanford, Sam Wineburg, calls this method "lateral reading," which emphasizes understanding the gestalt by pursuing multiple searches in parallel.[3]

Second, it often helps to simplify the text into a form that I understand. That is, I go sentence by sentence (or paragraph by paragraph) rewriting the article in language that I can comprehend. This is a bit slow, but it frequently really helps reduce complicated language into something you can understand. Don't be intimidated by complex language. Be a bold reader!

9. *They check their work.* Checking your research work seems obvious, but I often see people who do NOT check what they've found. One useful way to check your own work is to try to find a different answer. (This is a great way to overcome your own confirmation bias by trying to subvert your own research!) Just imagine that you're now working for an adversary, trying to prove that your other self's research is wrong and/or incompetent. How reliable are the sources that you're using as the basis for your findings? Is that set of sources really up to date? And can your imaginary adversary

reproduce your research? Checking your work is a great idea—both in sixth-grade math and especially the grown-up world of real online research.

10. *They validate the resources that they find and use.* That is, great researchers make it habit to check that their sources are credible, consistent with other sources, and accurate. Credible sources have a track record, give citations to other work (so you can check on their work), and even admit to errors. Beware the online resource that never admits to errors, and value ones that do. Newspapers and journals really should have a regular errata or update section where they correct blunders and mistakes that they've made. Similarly, professionals always check the date of publication; an encyclopedia from twenty years ago might have previously been accurate, but much of the contents will probably be out of date.

11. *They know when to stop searching.* It's really easy to just keep searching, following leads from one interesting topic to the next. That's why it's important to be clear about what question(s) you're trying to answer. It's OK to refine the question—but keep track of what you're trying to do. Stay on topic—or at least recognize that you're now working on a new and different question. That's OK too, but when you've found an answer to your research question, pause, check your work, and stop.

Attitudes of Great Researchers

The behaviors in this list are things that great researchers do in their searching. But there's also something else I've noticed about good researchers: it's not just the search skills they have, but the way that they approach these research questions. These people have an attitude. More precisely, they share a bunch of attitudes about their searching. And when I say "attitude," I don't just mean that they're punkish thugs with a swaggering attitude problem but rather that they have a particular mental outlook—a way of approaching their research that makes them productive, efficient, and accurate.

These common attitudes are as follows.

1. *Resilience.* One of the important differences between a merely OK researcher and a *great* one is that essential attitude of being resilient in the face of change. Since things change constantly, and since there's not much we can do about it, being able to compensate—that is, be resilient, work

around setbacks, and find another way to accomplish your goal—is a key defining attitude for great researchers. Psychologists have identified some of the factors that make you resilient: a positive attitude, a dose of optimism, an ability to limit your disappointment when things go wrong, and the ability to see failure as useful feedback.

Resilience is the attitude of being willing to take on something that seems tough with the confidence that you'll succeed and overcome any difficulties.

2. *Persistent.* If being resilient is the ability to gracefully recover from difficulties, being persistent is the attitude of trying over and over again. The best combination is being *persistently resilient.* The trick is to know when to stop being persistent because you're not getting anywhere.

Perhaps most of all, a persistent researcher is one who isn't frightened when they find something that's complicated. (I see this a lot. When someone is searching and finds some results that are slightly mathematical, they'll drop the whole thing as if it were covered in fire ants. Don't be that person.)

Persistence is also the attitude of being willing to take on something that seems tough with the confidence that you'll succeed and overcome any difficulties—and then, keep at it.

3. *Curious.* Of course, the way to know more stuff about a domain is to be curious about things in that area. Curiosity is the motivation to *find out more about something.* At the end of the day, this is what's going to matter. People who are curious about the world generally know more about it. (They're usually self-teachers who can't help themselves.) While curiosity is particularly strong in humans, we're not born knowing *how* to satisfy our curiosity; that's a skill set you really need to acquire. The important attitudinal difference is not just about *being* curious but also being able to do something about it. Curious searchers go deep and wide—they are typically persistent and resilient—and as a side effect of their curiosity, they generally just know more about the world.

4. *Abstract thinkers.* One of the basic problems I see in beginning researchers is the tendency to see only the specifics in what they're doing. *Thinking in abstractions* is the ability to work out not JUST the concrete details of the research topic but patterns in what you're searching for more generally as well.

For example, when doing some research about the eating habits of a particular kind of bird, you might find it helpful to search for the eating habits of that *genus* of bird rather than just the species. Abstraction thinking means that you're looking for the patterns that apply broadly instead of simply zeroing in on the details.

Penguins, such as the chinstrap penguin, typically eat krill, small fish, and other crustaceans. Another penguin, the king penguin, eats primarily fish, but not krill—and oddly enough, also small stones. Thinking abstractly about this makes you wonder if this is true for all penguins. Do they all consume stones? A curious searcher can quickly find out that this is true for more than just king penguins; it applies also to Adélie, African, gentoo, Magellanic, and yellow-eyed penguins. The "abstraction step" was to think about the whole category rather than just the one particular kind of penguin.

Another step of abstraction is to think about birds in general. Do other types of birds also swallow stones as they eat?[4]

5. *Learn from mistakes while searching.* When you're looking things up, stuff goes wrong all the time in the way that you're searching. It could be something simple—your search isn't specific enough—or it finds something that's *not quite right,* and you have to modify it. Usually, learning from your mistakes means identifying what went wrong in your use of the search tools and then remembering that lesson for the next time that you do this kind of research. In this way, people who search a lot learn not just from their successes but also from their mistakes. They think about what worked—and what didn't work—as a way to constantly improve their searching. Even better, when you learn from your mistakes, you can think about the errors of your way more *abstractly* and understand how that kind of error can be seen in other search cases.

6. *Constantly learn about the world.* The good-news, bad-news story about doing online research is that everything changes all the time. Not only do the tools you use to search change, but the underlying data you want to search changes constantly as new databases come online and older content vanishes. If you're constantly learning as you search, you're learning a good deal about the topic along the way. Consequently, an important attitude to have is the willingness (and commitment) to stay up to date on your topic as things change. Find a way to make *learning* part of your daily (or at least

weekly) routine. This is true of the internet age in general, but is especially true if you want to keep being a good researcher.

7. *Focus*. One of the things I've noticed in my own studies is that the closer you look at something (almost *anything*), the more you'll discover that darn near *everything* is interesting if you take the time to look carefully. You just have to appreciate the interestingness that's all around you! Naturally, with all that online information easily accessible, it's all too easy to get lost and spend your whole day in an interesting, unquestionably fascinating yet useless rathole. A key to success is the ability to keep your focus in mind, and not get lost in all the fascinating and sparkly bits. (This is a bit of the opposite of being curious; but if curiosity expands your mind, it's focus that brings the results home so you can get something done at the end of the day.) And this is the reason I write down my research questions: to remind myself of what it is I'm researching. It's a way to externalize your attention to the target of your research.

Rules of Thumb for Asking Good Questions

We've seen some common behaviors and some of the attitudes of great researchers. In addition to all this, there's the skill of asking good research questions to begin with. You've probably heard that "there are no bad questions," but there are definitely better and worse questions. As an example, here's an almost-useless question that you can ask, even though you shouldn't expect a good answer: *Who was the best baseball player of all time?* You can ask it, but be prepared for a long discussion about what "best" means in baseball.

Here's my list of the ways I see great researchers asking good questions.

1. *Think to yourself, What else could this mean?* In other words, how else could someone interpret your question? Simple questions often come with a lot of cultural baggage built in to the framing of the question. Something like "When is the first train on Monday?" might vary depending on whether or not Monday is a holiday. For a deeper research question—something like "What did Alexander Hamilton accomplish in his lifetime?"—the answer will vary a great deal depending on who's asking it and who's expecting the answer. A description of Hamilton's life given by a historian will be different than that given by fan of the Broadway musical *Hamilton*. This is true

even for something as simple as a boiled egg. The answer to "How should I boil an egg?" will vary on the details of what you mean. Are you a poached lightly kind of person or more of a hard-boiled fan? The details of what a question really means really matter.

2. *Does this question work for everywhere on earth or just near where I am?* Remember that easy questions often assume a particular location. The answer to "What kind of energy efficient light bulb should I buy?" will change a lot if you're in Europe, India, or the United States. This rule of thumb lifts you up from your localized thinking and into a different frame of reference—one that helps to clarify the way that your research question can be asked. It's the geolocation version of the previous question.

3. *Constantly fact-check*. Get into the habit of constantly asking if this thing that you're reading (or hearing or watching) is internally consistent (for instance, whether the numbers add up) and consistent with what you already know.

I find myself noticing things, dates, people, and places. I'm constantly asking the question, "Does what they're saying fit in with everything else I know?" I'm also asking, "Is this consistent with itself?" You'd be surprised at the number of times things just don't add up—I mean that literally (such as when the percentages of things don't add to 100 percent) and figuratively (as when the basic facts and assumptions in the piece are just plain wrong; I read an article the other day that had the population of the United States just *wrong*. Sometimes you just have to know some basic facts about the world, such as in 2018, the population of the United States is around 325 million, or knowing that sound doesn't travel in space (so you can't hear a spaceship blowing up, no matter how often you've heard it in science fiction movies), or knowing that Shakespeare lived during the reign of Queen Elizabeth I (which is why he wrote using an *Elizabethan* voice).

This back-of-the-brain, constant fact-checking becomes important when you're asking a question. If you know that around 17 percent of the US population is Hispanic in 2018, asking a question about the "100 million Hispanic people in the United States" is seriously off base to start. (That is, 17 percent of 325 million is about 55 million, so that frame is off by almost a factor of two.)

Whenever I read a news item or see a story reported on video, I find myself quietly adding up the statistics given (do they add up to 100 percent?) and cross checking the various pieces of data as presented. These kinds of sanity checks let you spot all kinds of errors in data, even when presented on well-meaning news channels.[5]

4. *When your research question covers a long period of time, consider making a timeline to keep track of everything.* Many questions can cover a large number of events or extend over a long period of time. (Take this sample research question with lots of actions over time: "What were the events that led to the collapse of the energy company Enron in 2001?") Great researchers not only fact-check constantly but also try to put events into a kind of timeline as a way of both keeping track of events and checking to see if the events make sense. This kind of timeline tracking not only helps you understand what happened in time but also *why* particular events happened as they did.

And in particular, having a timeline helps you to see what kinds of questions you should be asking, if only to fill in the gaps in the story line or if you don't understand why an event did NOT happen. Even a simple timeline makes these kinds of relationships apparent.

5. *Look for the context that gives meaning to information.* As we saw above, creating "context" is something that great researchers do all the time. When you're framing your question, it's useful to think about what *other* information will help you understand what you seek. That is, a great question is one that comes along with other questions in parallel for every question you ask, you also should be thinking about how to find the context. You're not just asking *what* is true but also *what else you need to know*.

20 The Future of Online Search: Why the Research Skills You Learn Today Will Continue to Be Useful in the Future

Where will online research be in five to ten years and what skills do you need to know now? Will those skills still be useful in ten years? Answer: there seems to be no end to the need for basic knowledge. No matter how skilled the online tools are, we'll still need to understand what's going on.

When I started this book, it was based on the premise that people are curious and want to know more about the world. Along the way, I've investigated many different curious questions, and learned how to do this kind of online research using a variety of resources and methods. We're living in a time of fast technology development and change. So an obvious question at this point might be, *How well will those skills last?*

After all, what's the point of learning all these skills and somewhat-hidden repositories of information? You might believe that in the future, Google will just give you the answer in response to your questions. So what will change over the next couple years? What skills are worth learning and practicing?[1] (To put it figuratively, what should you learn that won't become as useless as a screen door on a submarine?)

I've lived long enough to remember the world pre-Google. I know that might border on the inconceivable, but it's true; I was born before the internet, email, cell phone, or Google. Over the course of my life, doing online research has gone through astonishing change; the omnipresence of fast online search has changed everything—from writing a PhD thesis to settling questions raised in pubs over beer. (For one technology to make THAT broad of a change is truly phenomenal.)

This is a question we can think about by looking back for historical precedent.

If you cast your mind back to 1993, the first real search engine was Excite, which opened for business in that year. There were a few small search tools before then (though not quite full search engines in the way that we think about them now)—systems like Archie, Veronica, and the World-Wide-Web-Wanderer—but they were all fairly limited in what they could do. Just as important, the web wasn't that large. Online content was mostly created by hobbyists and scientists. There just wasn't that much good stuff to search.

Then the tide shifted; more content started becoming available online, and with real content came the realization that people really DID want to discover what that information had to tell them. And with that shift came the importance of search engines.

There were lots of different kinds of search engines early on—just after Excite came Lycos, Hotbot, and Alta Vista—and all were pretty good, but when Google search technology came online, it was clear that they were using hopelessly inadequate methods, and the research world had gone through an informational seismic event. Fairly quickly, using Google to search became the default online research method. Microsoft had its "Live" search engine, which was later updated by Bing. In all cases, modern search engines now use more up-to-date algorithms that take many factors into account when rank ordering the results of a search.

Early on, though, you could see that the world wasn't going to be as simple as what Larry Page and Sergey Brin had hoped for; there really was no way they were going to build a system that would crawl and index "all the world's knowledge," and then offer it up through a single interface. The divisions started early: Google Scholar broke off the "scholarly literature" part, and offered its own user interface and index. (It took another ten years before Google Scholar content was available through regular Google searches.) Now, Google's Images, Maps, Patents, News, Shopping, and Books all have their own access points, and—crucially—their own sharing of content with the main search engine.

For instance, the ability to search by image on Google was announced in 2011, and it is a completely remarkable tool that changes the way we search and how we think about images in general.

But you can't search Google Images for images from Google Street View. And contrariwise, you can't use regular Google web search to find any of the images in the Google Cultural Institute. (I mean, there is a way, and I can show you how, but you have to know how to do it; it's not obvious.)

Search engines try to "blend" all the information silos together by showing a search results page that has the best-of data from each siloed kind-of-content area. Search for [Disneyland], and you'll see images, news, and reviews all collected together. But there are many more kinds of information that could be shown. What about 3D models of the Sleeping Beauty Castle? What of videos made in the park? You can find them, but you have to search on a video site (such as YouTube), and you can't search by geographic location—at least not yet.

My point is that we, as an online culture, are rapidly moving into the information retrieval past; instead of a perfectly flat and universally accessible universe of information resources, the world of information remains obstinately clumpy and segregated. We once had a dream that a single system would integrate all our information resources into a single master repository. The holy grail was that we'd only need to search in one place—in one master catalog, as the one place to find everything. There would be one ring to rule them all, and one index to find them.

In retrospect, we should have known that wouldn't work; there are too many resources that are held (i.e., owned) by people who either don't want the information out there or want to make money from giving access to the information. You can't blame them—it's a way to capitalize on their creative work—but it also means that the information world won't ever provide even simple access to all content.

In the old days, you had to know what resources were available in order to do your research. And that's true once again—except that now we have a great tool (search engines) that will help you *find* those resources and then let you go down each of the rabbit holes in turn. That is, only if you know how, and if you know what tools, methods, and buckets of content exist. It might not occur to you to search for the data from the census of the United States (but you can at census.gov), or that you could possibly find your grandparents' immigration records (as I did), unless you first know that such a search would tap into online data that's been assembled, curated, and put online for your search.

By the same token, though, we now also have tools that can help us look at information in new and sometimes previously unthinkable ways. Again, IF you know how. Here are a few examples.

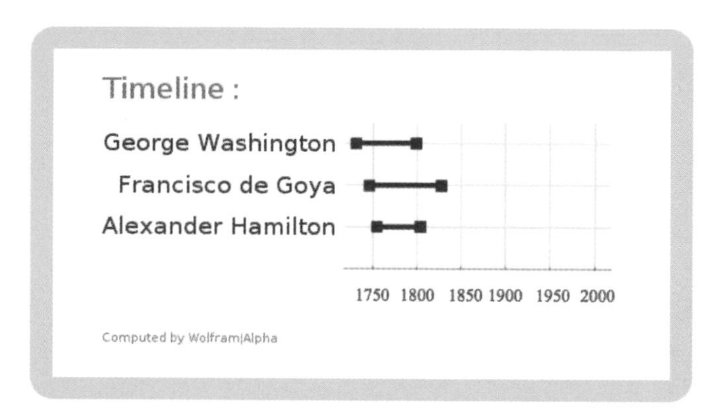

Figure 20.1
Creating a timeline is fairly simple, if you know which tool to use. A simple query on Wolfram|Alpha gives a nice timeline of the life spans of three famous people of the eighteenth century.
Credit: Wolfram|Alpha

Want to see a chart of the lifetimes of George Washington, Alexander Hamilton, and Thomas Jefferson? That's easy! Use the Wolfram|Alpha search engine and give the query:[2]

[George Washington, Alexander Hamilton, Francisco de Goya]

In response, Wolfram|Alpha creates a timeline with each person's dates, side by side, showing their lives with some context. Of course, you can add in other people to see how they relate to your topic of interest.

In a couple seconds, you'll have a really nice timeline diagram showing you when they all lived.[3]

Or want to get the elevation profile of an east-to-west path over the Himalayas? Use Google Earth, draw a path, and then click on "Elevation profile."

This is a fantastic tool, but you need to know about it. It's even better if you know the skill of how to find such a system to give you the data in the form you need.

For example, you might need to know that in the United States, the .gov top-level domain refers to government sites (and you need to know that it's .go in other countries). Likewise, if you're looking for reliable statistical data, you should know that there are a number of places where you can

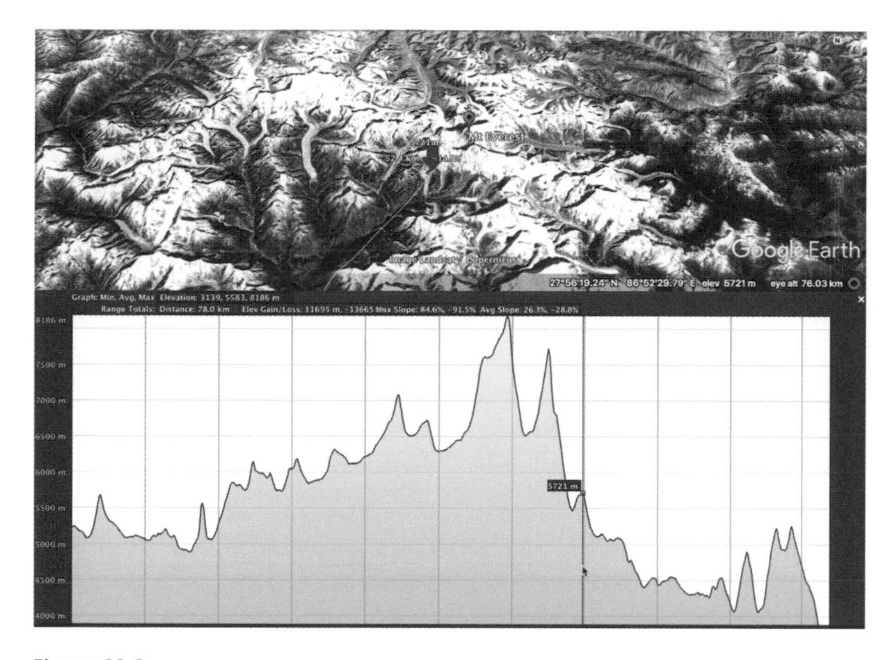

Figure 20.2
If you know what various tools can do, it's straightforward to find the answers to complex questions, such as making this vertical profile across the top of Mount Everest to get a sense for what the topography of the Himalayas is like.
Image © Google

find such tables; the Bureau of Labor Statistics has vast quantities of data, as does census.gov and the German Research Foundation's re3data.org site.[4] Google has its own data collection at the Google Public Data Explorer, and Amazon maintains its Open Data Registry.[5]

Overall, the web has made more information resources *closer* at hand. That is, what you seek is faster to find and quicker to gain access. Here I mean "closer" in the literal sense of when a book is physically closer in the library, but also in the figurative sense of when it's faster to get to the text, just as if it were closer to you. Lots of research shows that "closer" resources are understood more deeply and used more frequently by people in libraries and via online resources.[6] Putting things far away is death to that text, as is making something harder or slower to find.

This also has given us the illusion that anything that's off-line (or more than two minutes away in time as distance) isn't useful, or at least isn't used as much. This is certainly the case for online resources, where a delay of as little as 0.1 seconds can make a substantial difference in the likelihood of access and reading.[7]

As our ability to do online research has grown, we've come to realize five deep truths.

1. *There's more stuff online than you might imagine or know about—but it's not all uniformly accessible.* One of the surprises (and delights) of my work has been learning about the vast quantities of information that's available on an immense variety of topics. Want to learn about different diseases of abalone (fungal? viral? bacterial?)? It's simple to do. How about dressage style of the sixteenth century?[8] Sure. Not a problem. Want to see nineteenth-century records from Ellis Island immigration? That's simple too. There is just an amazing amount of content online.

In other words, the kinds of research that used to be a huge hassle are now fairly straightforward as more content becomes available. This stuff is either "born digital" content, or becomes available as older content is scanned and indexed. That's how you can learn about sixteenth-century Italian dressage; the Google Books project scanned and then indexed the original text.

One of the consequences of this is that new content is popping up all the time. New websites appear minute by minute (there were 1.3 million new website domain names registered in the first three months of 2017, up 3.7 percent from 2016).[9] New online applications let us see content in new ways (such as the InformationIsBeautiful.net site, which specializes in this kind of seeing-in-new-ways data, or Google Earth VR, which lets you fly through a 3D model of the Earth, including many city building landscapes).[10]

And the side effect of *that* is that you can't really keep up to date on what's newest and greatest in your field. Instead, you'll have to keep your eyes open and periodically search for new content in your area of interest. We have shifted into an age of information triage—separating out what's useful and important from the stuff that isn't.

Yet all this easy availability isn't uniform for all new content. Often that information is in silos that are owned by someone who wants to make some money off that information. That's a fine thing to do, but sometimes journals, newspapers, and other publications are behind paywalls; the content is frequently indexed and therefore searchable, but you can't get to it without paying.

2. *There's more different KINDS of stuff than you might have thought about.* Videos, virtual reality environments, 3D models, data sets, spreadsheets, presentations, blogs, vlogs, and 3D-printable objects—the number of different *kinds* of online content is astounding, and the number of different kinds of content continues to grow. This means not only does a literate person need to have a comprehensive understanding of the areas of information that's available but also an understanding of the different ways that information can be stored, indexed, accessed, and read.

What makes this complex is that there's no single authoritative news feed or information stream that will tell you about all the new types of data objects (and tools to use them).

In practice, great online researchers usually follow a number of online newsletters (to track the introduction of new content types, along with collections that pull them together). Given the number of products that come and go, I tend to only look at new data resources once there's a large enough collection of them to do a scan and see if it's worth me spending my time to evaluate them. If you track down every new introduction, you'll never get anything done.

3. *Online information often doesn't come with much context.* You have to supply your own or know how to find it. Just as important, you have to understand—DEEPLY understand—that context is just as critical as the information you seek. (As an example, it's not much use to know that the French and Indian War was held in 1754–1763—unless you also understand that it was part of a much larger global war, the Seven Years' War, which involved not just the nascent United States but also Great Britain, France, Germany, Spain, Russia, and Canada.)

Naturally, long-form texts (such as books and longer magazine or journal articles) are great at providing context; that's often why they're long. But in other kinds of research where the hits tend to be shorter and decontextualized, the job of finding context is left up to you, the researcher.

4. *It's also easy to find fake, spurious, or incorrect information as well.* Technology has made it easy to publish content on the web. That's great; the world now has more published information than ever, on a wider variety of topics

than ever. Just as it's simple for serious archivists, publishers, and librarians to add to the world's knowledge, however, it's also simple for those who are uncaring, or downright uninformed or malicious, to add content that will show up in your online searches. (One of my *favorite* bizarre stories along this line is the former NASA engineer who claims massive alien motherships are lurking inside the rings of Saturn, spawning UFOs that are plaguing the solar system.) You can find gems of content that are filled with insight and brilliance with your searching, but you can also find bullshit just as easily. It's all online and it's all searchable, but this means that you, the researcher, need to understand the author's motivations as well. You need to follow (at least a little bit) the big trends in publishing. Are false and misleading news articles popular this year? You need to know that, and what signs you should know in order to pick up the fake news.

5. *Nothing lasts forever, even online resources.* Although people commonly believe that "nothing on the internet is ever deleted," the truth is that researchers run into missing content all the time. When clicking on a link gives you a "404" error, that means that the target page of the link was moved or deleted. This happens more frequently than you might like. As a great researcher, you need to know how to be resilient in the face of vanishing content. When it happens (and it will happen to you), you'll need to search again for the missing content. You can often do this by searching by the page title (use the page title of the missing page with the **intitle**: search operator) or the URL of the old page (use the **inurl**: operator), or by double quoting key text that was in the document.

A skilled online researcher needs to know not just that a kind of content exists but also where it might be cached after the hosting website has vanished into the ether. You can use **cached**: as a Google search operator to retrieve the previous version of a web page (for up to six months after it's gone missing). And when things really go missing, you can often use Internet Archive to find missing content.[11]

In the future, this means that you'll continue to need these skills. Although they might well change in form over the next decade, the core ideas—how to narrow a search, how to find repositories of content, how to triangulate information from multiple sources, and how to recover lost or deleted content—will always be core skills.

These five truths suggest that online researchers need to know MORE about the information landscape, not less.

A caution about checklists: every so often you'll find teaching aids that say "in order to verify/validate information, do this," with a list of things to check off. The problem is, this "list of things to check" goes out of date quickly. A fundamental property of the online world is that things change constantly—the moment you put up a content quality checklist guide with rules like "check for the presence of a copyright date," sketchy online publishers figure it out and suddenly, copyright dates are everywhere, making the checklist less useful.

What Will the Future of Online Research Look Like?

As much as I'd like to lay out a clear road map for the future of online content along with search systems to help with that, that's a bit treacherous. Surely our ability to search will improve as algorithms and tools get better, but exactly how and in what ways it will change are tricky questions. That's my cautionary note; now let's look at some trends that seem inevitable over the next few years.

Mobile searching: The global trend toward more searches happening on phones is inexorable. The trade-off for searchers doing detailed research tasks is that the mobile devices have small screens. That makes it a bit more difficult to read large documents with lots of text. It's also harder to do lateral searching, using multiple tabs or multiple windows to explore comprehensively, or to save the state of your searches. On the other hand, you've got an amazing resource with you all the time. And because you're searching in a specific location, this gives you the opportunity to do local searches (for instance, queries like [coffee near me] works really well when you're on your mobile device).

Spoken search: Along with more searches on mobile devices comes an increasing trend to speaking your queries aloud. This is true in countries with high levels of literacy, but it is a special boon in countries with lower levels of reading and writing. Voice recognition for queries that works with text to speech is a godsend in such places. But even in high-literacy countries, spoken searches will continue to become an increasing part of our research culture.

Of course, spoken output isn't the only way to go with mobile devices. The results of search are often best seen and not heard. And while I don't expect a list of link results to disappear, the results of searches will become more media savvy and give visual resources when that's a better approach.

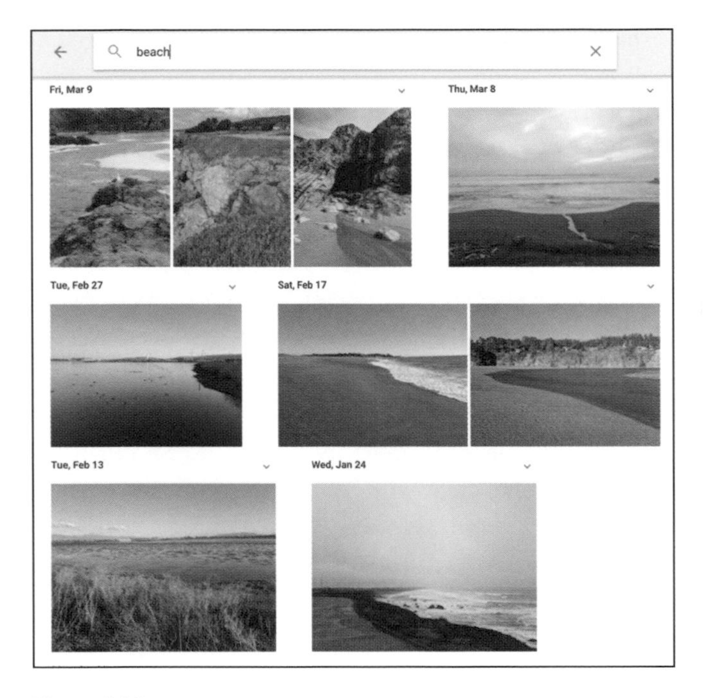

Figure 20.3
You can now search your personal photo collection for photos with different kinds of objects (in this case, a beach) within them. Note that none of these photos have the word "beach" attached to them; the AI system recognizes them as beaches.
Credit: Daniel M. Russell

More artificial-intelligence-based help: Artificially intelligent (AI) systems will change many of our expectations about how online search operates. You can already see this in your personal photos on Google. Searching your personal photos for something like [beach] produces a set of photos of beaches; the AI boost here is that none of these photos have the word "beach" in them—the AI system is attaching labels of things in the photo, letting me later search across my photos.

Right now this works well for things like "cat," "dog," "beach," "ship," and so on. But you can see that soon we'll be able to search for pictures with queries like [two golden retrievers playing with babies] and it will just work.

AI also powers entity recognition, which is the ability to identify all the variations on a named entity (e.g., "Dan Russell," "Daniel M. Russell," "Dr. Russell from Google," and "that guy from Google who works on search named Dan, or something …"), which will improve the ability to search for entity names, even if you don't know all the variations on that theme. Entity detection will also allow you to ask fairly general questions that depend on a particular kind of entity being listed in the target document. Imagine a query such as [California <technology-company> green energy] where the phrase in angle brackets defines a kind of entity that you want in your results—in this case, any technology company that is in California and working on green energy.

Question answering: Of course, one of the biggest targets of AI is the ability to process natural language well enough to answer questions and potentially have a bit of conversation about your research topic. As it is now, Google and Bing can both handle fairly straightforward queries written in question form. Questions such as [How many countries are in North America?] and [Who was the prime minister of the United Kingdom in 1980?] can be handled easily. You can see the answers starting to improve as both Google and Bing now provide a bit of context in the answers. In Google's answer to [What is the population of India?], you see the direct answer, but also a chart comparing its population with other similar countries and a panel on the right side with related statistics. (See figure 20.4.)

In essence, questions like this are translated into regular queries and handled in the normal way. But you can see where this is going with improved natural language. Search engines are starting to be able to answer follow-up questions that correctly handle pronouns. You can ask Google, [How tall is President Obama?], and then follow up that question with [How tall is his wife?] and then [How old are their children?]. While this is relatively straightforward, the direction is clear: search interactions can become much more like a conversation. Simple queries will always be with us, but the ability to ask questions and then follow up in the conversation will make much research far simpler.

More social: Social media is a big player in online content, if only for the powerful effect that recommendations have on people. When you're doing online research, your social network (and implicitly, all the social media systems and tools) can be an incredibly valuable ally in your research. The players may change over time (anyone remember the MySpace social media site?), yet it's worth cultivating your personal network of friends in the know to ask them for help. In my studies of really extraordinarily skilled researchers, one striking thing I noticed was that they all had (and used) deep and broad Rolodexes, or what would today be called "social contact lists."

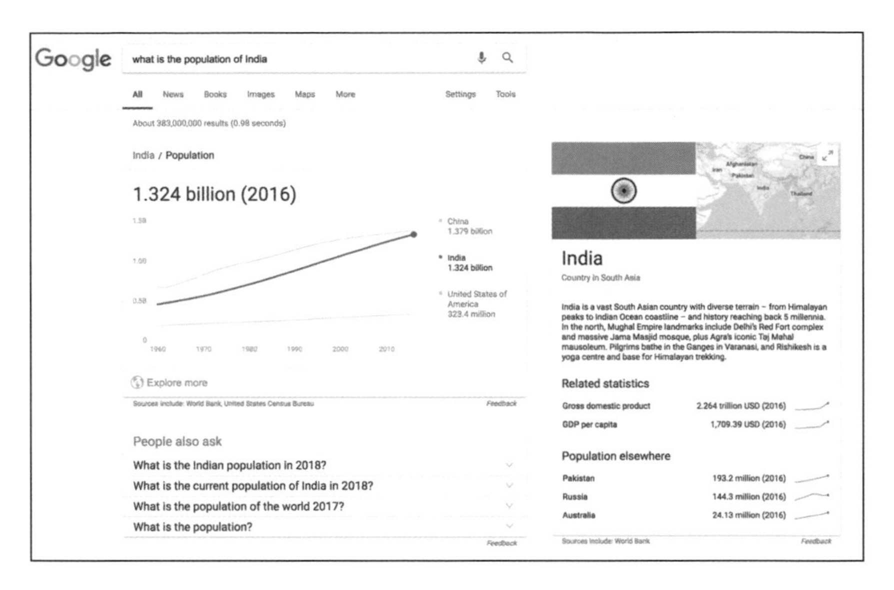

Figure 20.4
Search engines can handle queries that are written out as questions. The ability to do this will continue to improve. The answer shown here provides some nice context information, such as the population graph that compares India's population with that of the United States and China—a comparison that makes the number of Indians easier to understand.
Credit: Google and the Google logo are registered trademarks of Google Inc., used with permission

More assistance: A huge trend in the industry is toward more and more online assistance that goes well beyond research. Mostly, this is a trend toward helping you get everyday tasks done quickly and accurately, but you can see a day when online assistance systems improve your ability to manage multiple research tasks or data sets. One can also safely predict more intelligent agents to provide information-finding aids. Imagine a chatbot reference librarian who knows about the catalog of every library in the world and every reference book out there in their collections.

Content continues to grow in quantity and kind: There's every reason to believe that content will continue to grow, both in the amount of online content and the different kinds of content. A perpetual challenge as new kinds of resources become available will be how well they can be indexed. Although it doesn't make much sense to create a new kind of online content that can't be found by search engines or online catalogs, it happens every day, leaving unhappy researchers to manually track down the content without the aid of an online index. One side effect is that media literacy will be an ever-evolving concept. How media is used will continue to change, and as new kinds of media evolve, so too will the way that they're used.

Content continues to splinter: It's not much of a prediction to point out the obvious, which is that copyrights will continue to operate, and new content providers will arise. The side effect of these two forces is that searchable content will continue to be in unsearchable silos. The good news here is that there are also shifts toward increasingly open access publication. But as the publishing industry continues to search for an effective funding model, we, the researchers, should expect information to still be in repositories, some of which are open, and some of which are not. Some will require payment to read their content, and others will be free and open to the public.

Then there's regulation—the big unknown in the content space, but a highly contested area as publishers look for protection (and payment), while searchers look for answers to research questions. The only thing that's predictable here is that more regulation will happen, making things increasingly complicated.[12]

I worry about the future of online research, but my deep belief is that the ecosystem of online content and search over that content is just too useful, too handy, and too important to suddenly go away. The one thing I am certain of is that the landscape of content will change constantly. Sites, data sets, maps, images, recordings, and videos—the whole lot of it—will come and go. Sometimes that change will be driven by deep-seated, honest desires to improve knowledge in the world, sometimes the change will be driven by people who are trying to manipulate information for personal gain, and sometimes the change will happen by accident, when a company (and the information associated with it) goes bankrupt or is absorbed into another company that doesn't see the value in having the information publicly available.

In other words, the information universe will change shape and content, just as it always has. The big difference is that now there are larger amounts of information and change happens far more quickly.

Likewise, the tools that we use to access, manipulate, and search for that information will continue to change as well. As we've seen, kinds of content will come online, and old kinds will disappear. There will be wholly new ways to see information and interact with it. Probably some of those new kinds of content and new tools will vanish as well, like tears in rain.[13]

But the underlying skills, attitudes, and behaviors you have learned here will continue to be valuable. Knowing how to evaluate the quality of content, and understanding how to go from one resource to another—these are powerful and eternally useful skills. In fact, if anything, these search and research skills will become more valuable, not less so, as things change. The range and diversity of online resources will continue to expand—as will the need to understand them, and discern their quality and accuracy.

I started this book with the idea that you too could learn how to ask (and answer) your odd and peculiarly interesting questions. You've seen how I do it, but this is just the beginning; there's so much more to learn.

Go forth, and start asking—and answering—your own curious questions.

Notes

Chapter 1

1. Quoted in James Boswell's *Life of Johnson*, 1791, bit.LY/TJOS-1-1.

2. Ibid.

3. For readers on the younger side, an *encyclopedia* was a huge set of books with many, many articles on different topics, written by experts, arranged in alphabetic order by the subject that they covered. You had to learn the skill of looking for multiple articles if you wanted to get a complete, well-rounded understanding of a topic like "England." It would be handy if you also read other articles like "United Kingdom" or "Queen Elizabeth." The better encyclopedias had little hints—links really—to other articles in the encyclopedia. But you had to actually pull out the other volumes (and we're talking about big books) to read them all. Now, of course, you'd just click on the link to go read the other articles instantly. You can't believe how much faster clicking is rather than pulling out all the volumes. Remember, you have to put them all back into the bookcase ... in alphabetic order!

4. A Boolean query is one that uses some special operators—AND, OR, and NOT—to create its query. Note that neither AND or NOT are considered special operators in modern search engines.

Chapter 2

1. How do I know this isn't Cairo? Because you can get a quick look around the Cairo cityscape by doing a Google Images search for [Cairo]. That will give you a quick set of Cairo cityscape pictures. When you look at them, you'll find that almost

every picture has the minaret of a mosque in it as well as some Arabic writing. This one has neither.

2. Google Earth is a free application that you can download from Google that gives you the ability to fly anywhere in the world, and take a look at the terrain, images, and if you want, 3D models of the buildings in that area. bit.LY/TJOS-2-1.

Chapter 3

1. Here's a sample link to an article from the *New York Times*, "Warning on Gas in Cameroon Lake," 1991: bit.LY/TJOS-3-1.

2. Here's a link to a map of Lake Nyos: bit.LY/TJOS-3-2. Nyos is the brownish circular lake in the middle of this view on Google Maps.

3. George W. Kling et al., "The 1986 Lake Nyos Gas Disaster in Cameroon, West Africa," *Science* 236 (1987): 169, bit.LY/TJOS-3-3.

4. Youxue Zhang, "Dynamics of CO_2-Driven Lake Eruptions," *Nature* 379 (1996): 57–59, bit.LY/TJOS-3-5.

Chapter 4

1. bit.LY/TJOS-4-1; https://www.snopes.com/fact-check/fuel-icon-foolery/.

2. bit.LY/TJOS-4-2; https://www.phrases.org.uk/meanings/at-one-fell-swoop.html.

3. bit.LY/TJOS-4-3; http://www.imperium-romana.org/uploads/5/9/3/3/5933147/scientificamerican1073-34.pdf. This is a remarkable article about how powerful a simple handheld sling can be.

4. bit.LY/TJOS-4-4; http://digitalcommons.unl.edu/nebanthro/169/.

5. bit.LY/TJOS-4-5; https://books.google.com/books?id=oICRAwAAQBAJ&printsec=frontcover&dq=David+and+Goliath:+Underdogs,+Misfits,+and+the+Art+of+Battling+Giants&hl=en&sa=X&ved=0ahUKEwjq1b6_pJraAhXFqYMKHUnSBmEQ6AEIKTAA#v=onepage&q=sling&f=false.

Chapter 5

1. bit.LY/TJOS-5-1; https://www.accessdata.fda.gov/scripts/plantox/detail.cfm?id=11976.

2. bit.LY/TJOS-5-2; https://books.google.com/books?id=78FC6xQcNP8C&printsec=frontcover&dq=A+field+guide+to+Pacific+states+wildflowers&hl=en&sa=X&ved=0ahUKEwjJq-aY0pnbAhUEjlkKHY5sCp0Q6AEIKTAA#v=onepage&q=A%20field%20guide%20to%20Pacific%20states%20wildflowers&f=false.

3. This reminds me a bit of the debate over whether or not Pluto is a planet, or merely a lightweight *dwarf planet*. For a lively summary of this planetary naming kerfuffle, see bit.LY/TJOS-5-3; https://en.wikipedia.org/wiki/Pluto.

4. bit.LY/TJOS-5-4; https://baynature.org/article/the-scent-of-summer/.

Chapter 6

1. bit.LY/TJOS-6-1; https://www.cdc.gov/nchs/products/nvsr.htm.

2. bit.LY/TJOS-6-2; https://www.cdc.gov/nchs/products/nvsr.htm.

3. bit.LY/TJOS-6-3; https://unstats.un.org/unsd/demographic/products/vitstats/serATab3.pdf.

4. bit.LY/TJOS-6-4; https://www.ncbi.nlm.nih.gov/books/NBK453383/figure/mortality.f3/?report=objectonly.

5. bit.LY/TJOS-6-5 and bit.LY/TJOS-6-6; https://www.cdc.gov/injury/images/lc-charts/leading_causes_of_death_age_group_2016_1056w814h.gif and https://www.cdc.gov/injury/wisqars/LeadingCauses.html.

Chapter 7

1. Yes, I know you can edit the articles, changing their content, but in practice, this is harder than you might think. Have you actually tried to edit an article? It takes a special kind of person to work within the Wikipedia system. There's a fair bit of evidence that the overall quality of Wikipedia articles is quite high. You can see a comprehensive analysis of many studies in the article "The Sum of All Human Knowledge," in which Wikipedia is shown to favorably compare with the long-established Encyclopedia Britannica. Mustafa Mesgari et al., "A Systematic Review of Scholarly Research on the Content of Wikipedia," *Journal of the Association for Information Science and Technology* 66, no. 2 (2015): 219–245. bit.LY/TJOS-7-1; http://orbit.dtu.dk/files/103083646/WikiLit_Content_open_access_version.pdf. Alternatively, you can read Wikipedia's own assessment of its accuracy in the article "Reliability of Wikipedia" (bit.LY/TJOS-7-2; https://en.wikipedia.org/wiki/Reliability_of_Wikipedia). It's nicely self-referential, but also a decent analysis of what works and what doesn't. There are many such studies, all of which tell us that the quality of Wikipedia articles is remarkably high.

2. For the Wikipedia entry about Leonardo's painting and corresponding one about Michelangelo's, see bit.LY/TJOS-7-3 and bit.LY/TJOS-7-4; https://en.wikipedia.org/wiki/The_Battle_of_Anghiari_(Leonardo_da_Vinci) and https://it.wikipedia.org/wiki/Battaglia_di_Cascina_(Michelangelo).

3. For a Wikipedia article about its "featured articles," which are of especially high quality in Wikipedia land, see bit.LY/TJOS-7-5; https://en.wikipedia.org/wiki/Wikipedia:Featured_articles.

4. bit.LY/TJOS-7-6; https://en.wikipedia.org/wiki/Battle_of_Cascina_(Michelangelo).

5. bit.LY/TJOS-7-7; https://en.wikipedia.org/wiki/Wikipedia:Featured_articles.

Chapter 8

1. bit.LY/TJOS-8-0. Many other geology journals agree on this. As this publication says, "Barrier Islands are always found along passive plate margins." That's another way of saying that they're geologically quiet, flat, and boring. http://www.sepmstrata.org/page.aspx?pageid=305.

2. bit.LY/TJOS-8-1; en.wikipedia.org/wiki/Plate_tectonics.

3. bit.LY/TJOS-8-3-1; https://web.archive.org/web/20150607181159/http://californiasislands.com:80/2010/09/29/east-coast-vs-west-coast/.

4. bit.LY/TJOS-8-4; https://clasticdetritus.com/.

5. bit.LY/TJOS-8-5; https://www.ravenmaps.com/48state-drainage-landform.html.

Chapter 9

1. bit.LY/TJOS-9-0; https://en.wikipedia.org/wiki/Mud%C3%A9jar.

2. Terry Ruscin, *Mission Memoirs* (San Diego: Sunbelt Publications, 1999), 167.

3. To learn how to use the pegman, see the "How to Do It' section later in this chapter.

4. bit.LY/TJOS-9-4; https://99percentinvisible.org/episode/the-fancy-shape/.

Chapter 10

1. bit.LY/TJOS-10-1; http://www.redlandsfortnightly.org/papers/schuiling08.htm.

2. bit.LY/TJOS-10-2; ftp://ftp.consrv.ca.gov/pub/dmg/pubs/cg/1970/23_02.pdf.

3. bit.LY/TJOS-10-3; https://www.chevron.com/about/history.

4. bit.LY/TJOS-10-4; https://en.wikipedia.org/wiki/Pico_Canyon_Oilfield.

5. bit.LY/TJOS-10-5; https://www.thefreelibrary.com/GHOSTS+OF+AN+ERA+MENTRYVILLE+IS+A+MONUMENT+TO+BOTH+THE+START+AND...-a0106274717.

6. bit.LY/TJOS-10-6; http://articles.latimes.com/1993-02-21/local/me-923_1_oil-drills.

7. bit.LY/TJOS-10-7; https://en.wikipedia.org/wiki/History_of_oil_in_California_through_1930.

8. bit.LY/TJOS-10-8; https://books.google.com/books?id=OPgjAAAAMAAJ&dq=%22North%20American%20and%20Middle%20Eastern%20Oil%20Fields%22&source=gbs_book_other_versions.

9. bit.LY/TJOS-10-9; http://ohp.parks.ca.gov/ListedResources/Detail/543. For a Google Maps look at Petrolia's location on the "Lost Coast" of Northern California, see bit.LY/TJOS-10-10; https://www.google.com/maps/place/Petrolia,+CA+95536/@40.3138589,-124.3319552,32291m/data=!3m1!1e3!4m5!3m4!1s0x54d444e04f3b0f65:0x62a91272c52f856e!8m2!3d40.3253487!4d-124.2861705?hl=en.

10. bit.LY/TJOS-10-11; https://archive.org/stream/boxsouthgeology00calirich#page/n705/mode/2up/search/Mattole.

11. bit.LY/TJOS-10-12; https://books.google.com/books?id=13POAAAAMAAJ&printsec=frontcover#v=onepage&q=mattole&f=false.

12. bit.LY/TJOS-10-13; https://books.google.com/books?ei=-u4dTYLpC47ksQPYvMXIAg&ct=result&sqi=2&id=K2BYAAAAMAAJ&dq=Early+California+Oil%3A+a+photographic+history%22+-+-1865+-+1940&focus=searchwithinvolume&q=Union+Mattole+Oil+company+.

13. Primarily Josiah Stanford, elder brother to Leland Stanford of railroad and Stanford University fame.

14. bit.LY/TJOS-10-15; ftp://ftp.consrv.ca.gov/pub/oil/history/History_of_Calif.pdf, Walter A. Stalder, "Oil and Gas Production: History in California," published in *California Oil World*, vol. 34. no. 21, pt. 2, Nov. 1941.

15. bit.LY/TJOS-10-17; https://mattolehistory.wordpress.com/2010/12/21/basic-timeline-of-mattole-history/.

16. bit.LY/TJOS-10-17-1; https://mattolehistory.files.wordpress.com/2010/12/1907gouldarticle2-humstandardsm.jpg.

17. bit.LY/TJOS-10-18; https://books.google.com/books?id=vMRMAAAAMAAJ&dq=discovery%20of%20oil%20in%20california%20history&pg=PA9#v=snippet&q=doheny&f=false.

18. If you have to say your book is reliable in the title, you probably should worry about its accuracy.

19. bit.LY/TJOS-10-20; https://en.wikipedia.org/wiki/California_oil_and_gas_industry.

20. bit.LY/TJOS-10-15; ftp://ftp.consrv.ca.gov/pub/oil/history/History_of_Calif.pdf.

21. bit.LY/TJOS-10-21; https://en.wikipedia.org/wiki/Chicxulub_crater.

Chapter 11

1. bit.LY/TJOS-11-1;https://www.sciencedaily.com/releases/2015/04/150407085256.htm.

2. bit.LY/TJOS-11-2; https://books.google.com/ngrams/graph?content=Delhi+belly&case_insensitive=on&year_start=1800&year_end=2000&corpus=15&smoothing=3&share=&direct_url=t4%3B%2CDelhi%20belly%3B%2Cc0%3B%2Cs0%3B%3BDelhi%20belly%3B%2Cc0%3B%3BDelhi%20Belly%3B%2Cc0.

3. bit.LY/TJOS-11-3; https://ehistory.osu.edu/exhibitions/cwsurgeon/cwsurgeon/medicalterms.

4. bit.LY/TJOS-11-4; https://books.google.com/books?id=3aEJZRIxjDAC&pg=PR20&dq=Civil+War+language+disease&hl=en&sa=X&ved=0ahUKEwjxk56TxbnNAhUJ7GMKHeXBDgEQ6AEIIzAB#v=onepage&q=flux&f=false.

5. bit.LY/TJOS-11-5; https://www.battlefields.org/learn/articles/civil-war-casualties.

6. bit.LY/TJOS-11-6; https://chroniclingamerica.loc.gov/lccn/sn85042460/1887-03-31/ed-1/seq-6/.

7. bit.LY/TJOS-11-7; https://www.esalen.org/page/esalen-hot-springs.

8. bit.LY/TJOS-11-8; https://chroniclingamerica.loc.gov/.

9. bit.LY/TJOS-11-12; https://chroniclingamerica.loc.gov/lccn/sn85066387/1895-07-09/ed-1/seq-11/.

10. bit.LY/TJOS-11-9; https://books.google.com/books?id=NqRdAAAAcAAJ&pg=PA55&dq=apoplexy&hl=en&sa=X&ved=0ahUKEwjJqb_f1bnNAhVOwmMKHSa3A6YQ6AEINzAF#v=onepage&q=apoplexy&f=false.

11. bit.LY/TJOS-11-10; https://books.google.com/books?id=xEJDAQAAMAAJ&pg=PA90&lpg=PA90&dq=john+dolland+apoplexy&source=bl&ots=mQJKikLjrG&sig=HbHxayz2FpFH7-dwRFE_yctAJ3c&hl=en&sa=X&ved=0ahUKEwjFmJXh3-_YAhVY4mMKHTXfC_gQ6AEIUjAG#v=onepage&q=john%20dolland%20apoplexy&f=true.

12. bit.LY/TJOS-11-11; https://ht.ac.uk/. Hosted by the University of Glasgow, the Historical Thesaurus of English is a historical record for many of the words in English, going all the way back to Anglo-Saxon times. It's also behind a paywall, but many libraries have access through their systems. This is one of the reasons that you want to have a public or academic library card!

13. Books.Google.com. See also: https://babel.hathitrust.org/.

Chapter 12

1. https://news.google.com/newspapers.

2. bit.LY/TJOS-12-1; http://wikimapia.org/#lang=en&lat=38.054461&lon=-122.201228&z=17&m=w&show=/26258105/The-wreck-of-the-Garden-City-Ferry.

3. bit.LY/TJOS-12-2; https://theprivatenaturalist.wordpress.com/2013/07/29/derelict-ferry-boat-of-carquinez-strait-the-garden-city/.

4. bit.LY/TJOS-12-3; https://news.google.com/newspapers?nid=1345&dat=19850314&id=9ANMAAAAIBAJ&sjid=svkDAAAAIBAJ&pg=6246,3112715.

5. bit.LY/TJOS-12-8; https://news.google.com/newspapers?nid=2245&dat=19830809&id=-ZozAAAAIBAJ&sjid=nDIHAAAAIBAJ&pg=5429,4511590.

6. bit.LY/TJOS-12-4; https://books.google.com/books?id=-jvQUdBIpNUC&printsec=frontcover&dq=Ferries+of+San+Francisco+Bay&hl=en&sa=X&ved=0ahUKEwjc9Lu3k_TaAhXrsVQKHST5D8wQ6AEIKjAA#v=onepage&q=%22garden%20city%22&f=false.

7. bit.LY/TJOS-12-5; https://books.google.com/books?id=BZxNdgQq5LgC&pg=PA118&dq=%22garden+city%22+ferry&hl=en&sa=X&ved=0ahUKEwjW_uq0p_naAhUs5oMKHU-6B5UQ6AEIWzAJ#v=onepage&q=%22garden%20city%22%20ferry&f=false.

8. bit.LY/TJOS-12-6; https://www.cocohistory.org/index.html.

9. bit.LY/TJOS-12-7; https://en.wikipedia.org/wiki/Exif.

Chapter 13

1. Whereas common names are frequently ambiguous. For instance, there are three different birds commonly called "bluebirds" in the United States alone: western, eastern, and mountain. They're all different.

2. bit.LY/TJOS-13-1; https://whyevolutionistrue.wordpress.com/2013/11/05/fly-with-ant-mimic-wings/.

3. Yes, I know there's no evidence that he ever said this pithy phrase. But it's commonly attributed to Asimov. That's why I wrote "purportedly said." See this brilliant piece of Asimovian quote investigation by the Quote Investigator, Garson O'Toole: bit.LY/TJOS-13-2; https://quoteinvestigator.com/2015/03/02/curcka-funny/.

4. bit.LY/TJOS-13-3; https://dotearth.blogs.nytimes.com/2013/11/04/survival-of-the-extraordinary-a-fly-with-ants-on-its-wings.

5. bit.LY/TJOS-13-4; http://www.biodiversityinfocus.com/blog/.

6. bit.LY/TJOS-13-5; http://www.biodiversityinfocus.com/blog/2013/11/06/ants-spiders-or-wishful-thinking/.

7. bit.LY/TJOS-13-6; https://books.google.co.uk/books?id=fIwoAAAAYAAJ&dq=Goniurellia+tridens&lr=&hl=en. While not heavy on plot, the book does a great review of all the known (at the time) fruit fly genera south of the US-Mexico border.

8. bit.LY/TJOS-13-4; http://www.biodiversityinfocus.com/blog/.

Chapter 14

1. For background on cemeteries as places for family visits, and why running through one isn't as crazy as it seems, see Keith Eggener, *Cemeteries* (New York: Library of Congress Visual Sourcebooks, 2010). This book has a lovely essay describing our changing attitudes about what's appropriate behavior in a cemetery, and how cemeteries led to the design of America's first public parks.

2. Laura Rice, *Maryland History in Prints, 1743–1900* (Baltimore: Maryland Historical Society, 2001), 43; bit.LY/TJOS-14-1-1; https://books.google.com/books?id=FwgSAQAAIAAJ&dq=Maryland+History+In+Prints+1743-1900&focus=searchwithinvolume&q=rum.

3. Patrick Richard Carstens and Timothy L. Sanford, *Searching for the Forgotten War—1812: United States of America* (Bloomington, IN: Xlibris, 2011); bit.LY/TJOS-14-2; https://books.google.com/books?id=SX8Y71-dQHkC&pg=PT638&lpg=PT638&dq=Tonnant&hl=en&sa=X&ved=0ahUKEwj4-trxwp7ZAhVo7oMKHSyWBSkQ6AEINTAC#v=onepage&q=Tonnant&f=false.

4. bit.LY/TJOS-14-3; https://books.google.com/books?id=tDFkAwAAQBAJ&printsec=frontcover&source=gbs_ge_summary_r&cad=0#v=onepage&q&f=false.

5. Interestingly, Key wrote a different set of lyrics for the same tune in 1805, but this earlier version was about the First Barbary War and its US Navy hero Stephen Decatur. In essence, Key was recycling an earlier idea, with some of the lyrics being partly reused in the "Defence of Fort M'Henry."

6. He actually said, "*Dans les champs de l'observation le hasard ne favorise que les esprits préparés.*" As recorded in his lecture at the inauguration for the Faculty of Sciences at the University of Lille, December 7, 1854.

7. Yes, *that* Aaron Burr, who, despite shooting Alexander Hamilton in their famous duel, went on to serve out the remainder of his term as vice president, all charges having been dropped. But he then went on to be accused of conspiracy to set up his own nation in the lands along the Ouachita River, in what is now Louisiana. Key was an assistant lawyer in this case.

8. bit.LY/TJOS-14-4; https://en.wikipedia.org/wiki/Philip_Barton_Key_II.

9. Teresa Sickles (nee Bagioli) was fifteen (or sixteen, depending on which account you read) when she married Dan Sickles. Together, she and Key II attended parties and were generally well known to Washington, DC, society as an unusual couple since neither seemed to be regularly in the company of their spouses.

10. bit.LY/TJOS-14-5; https://books.google.com/books?id=0kwwl-wbFjIC&printsec=frontcover&dq=American+Scoundrel:+The+life+of+the+notorious+Civil+War+General+Dan+Sickles&hl=en&sa=X&ved=0ahUKEwit6d7WtrjaAhVEslQKHeA3A90Q6AEIKTAA#v=onepage&q=American%20Scoundrel%3A%20The%20life%20of%20the%20notorious%20Civil%20War%20General%20Dan%20Sickles&f=false.

11. This is a variation on the so-called Pomodoro Technique of setting a timer to break up study sessions into small, timed segments, with breaks between each twenty-five-minute interval to help in assimilating what you've studied thus far. I use this idea, but tend to set my timer for one hour. When the timer goes off, I do that same collecting of what I've learned so far, summarizing my notes, and assimilating.

12. It's called the Matthew effect after the verse in the New Testament's Book of Matthew where it is written in 13:12 that "for whosoever hath, to him shall be given, and he shall have more abundance: but whosoever hath not, from him shall be taken away even that he hath." That is, the rich get richer, and the poor get poorer.

13. bit.LY/TJOS-14-1; https://en.wikipedia.org/wiki/Wikipedia:Wiki_Game. The key idea of the game is to start at a random Wikipedia page and then navigate, using only Wikipedia links, to some target page. The player with the fewest clicks from starting page to the target wins the game.

Chapter 15

1. bit.LY/TJOS-15-1; https://www.fs.fed.us/database/feis/plants/shrub/adefas/all.html.

2. bit.LY/TJOS-15-2; http://www.scparks.com/Portals/12/pdfs/QH_Plant_Life.pdf.

3. bit.LY/TJOS-15-3; http://www.californiachaparral.org/images/Torry_Bot_Halsey_Allelopathy.pdf.

4. bit.LY/TJOS-15-4; https://books.google.com/books?id=RYpbyHVuzh4C&printsec=frontcover&dq=Perspectives+on+plant+competition&hl=en&sa=X&ved=0ahUKEwiSpuGCgPTaAhXirlQKHe_5AhwQ6AEIKjAA#v=snippet&q=allelopathy&f=false.

5. Ibid.

6. Bruce Bartholomew, "Bare Zone between California Shrub and Grassland Communities: The Role of Animals," *Science* 170, no. 3963 (1970): 1210–1212. This is easy to find using Google Scholar. You should give it a try.

7. bit.LY/TJOS-15-5; https://baynature.org/article/landscape-shaped-fear/.

Chapter 16

1. bit.LY/TJOS-16-1; https://www.discoverlosangeles.com/blog/historical-timeline-los-angeles.

2. bit.LY/TJOS-16-2; http://www.lulu.com/shop/pamela-munro-and-julia-bogany-and-virginia-carmelo-and-mary-garcia/now-youre-speaking-our-language-large-print-edition/paperback/product-3777088.html.

3. bit.LY/TJOS-16-3; https://en.wikipedia.org/wiki/Point_Conception.

4. Ibid.

5. bit.LY/TJOS-16-4; https://books.google.com/books?id=-chO8R-i3PYC&pg=PA467&lpg=PA467&dq=point+conception+cabrillo&source=bl&ots=CK00RvSFkI&sig=9a0PVQnXO0nHYkOdw1gsHdG6Vrg&hl=en&sa=X&ved=0ahUKEwjKwNnsn4LWAhXHqlQKHQtDAH8Q6AEIcTAO#v=onepage&q=point%20conception%20cabri&f=false.

6. bit.LY/TJOS-16-5; https://www.budapest.com/city_guide/general_information/history.en.html.

7. bit.LY/TJOS-16-6. This in turn links to the Wikipedia article on the history of Tanzania. https://en.wikipedia.org/wiki/History_of_Tanzania.

8. bit.LY/TJOS-16-7. See, for example, Oxford Bibliographies, http://www.oxfordbibliographies.com/view/document/obo-9780199846733/obo-9780199846733-0023.xml.

9. bit.LY/TJOS-16-8; https://books.google.com/books?id=sj9mDQAAQBAJ&pg=PA105&dq=history+of+boma+congo&hl=en&sa=X&ved=0ahUKEwiijf-Ft4LWAhVriVQKHSfRAAkQ6AEILzAB#v=onepage&q=history%20of%20boma%20congo&f=false.

10. bit.LY/TJOS-16-9; http://geonames.nga.mil/namesgaz/.

11. bit.LY/TJOS-16-10; https://books.google.com/books?id=REtEXQNWq6MC&printsec=frontcover&dq=Historical+Gazetteer+of+the+United+States&hl=en&sa=X&ved=0ahUKEwjv4tP1vo_aAhVBU98KHWcvDhAQ6AEIKTAA#v=onepage&q=%22los%20angeles%22&f=false.

12. bit.LY/TJOS-16-11; https://en.wikipedia.org/wiki/Abyssinia.

13. That's really because *nobody* makes a gazetteer or place-name dictionary for fun. They do it out of a real need, driven by a real passion. Such people go to amazing lengths to make sure they get the information right!

Chapter 17

1. bit.LY/TJOS-17-1; https://reefguide.org/.

2. Search skill question: Why didn't I include the double quotes à la "stoplight parrotfish"? Because I figured that the bigram (the pair of adjacent words) *stoplight* and then *parrotfish* would be pretty common only as a fish reference. So I didn't bother.

3. bit.LY/TJOS-17-2; https://en.wikipedia.org/wiki/Parrotfish.

4. bit.LY/TJOS-17-3; https://www.newsweek.com/many-white-sand-beaches-are-basically-fish-poop-369494.

5. bit.LY/TJOS-17-4; https://link.springer.com/article/10.1007/s00227-010-1411-y#page-1.

6. P. Frydl and C. W. Stearn, "Rate of bioerosion by parrotfish in Barbados reef environments," *Journal of Sedimentary Research* 48, no. 4 (1978): 1149–1157. bit.LY/TJOS-17-10; http://citeseerx.ist.psu.edu/viewdoc/download?doi=10.1.1.900.85&rep=rep1&type=pdf.

7. bit.LY/TJOS-17-5-1; http://eol.org/pages/5214/details.

8. Many libraries offer an interlibrary loan service. This lets you borrow from a wide number of other libraries just as though it was in your local library. It usually includes collections from academic institutions as well as many other public libraries in your area.

9. H. V. Thurman and H. H. Webber, "Benthos on the Continental Shelf," in *Marine Biology* (Columbus, OH: Charles E. Merrill Publishing, 1984), 303–313.

10. bit.LY/TJOS-17-6; https://www.floridamuseum.ufl.edu/discover-fish/species-profiles/sparisoma-viride/.

11. bit.LY/TJOS-17-7; http://science.sciencemag.org/content/133/3446/98.

12. The ability to do unit conversions (e.g., meters to feet, or pounds per acre into kilograms per hectare) is one of the lesser-known but incredibly valuable features of Google search. You can even convert [300 cubits in feet] or [300 cubits in meters] depending on if you prefer feet, meters, or cubits as your distance measure.

13. bit.LY/TJOS-17-8; https://scholar.google.com/scholar?hl=en&as_sdt=0%2C5&q=%22Bioerosion+and+sediment+ingestion+by+the+Caribbean+parrotfish+Scarus+vetula+and+Sparisoma+viride%3A+implications+of+fish+size%2C+feeding+mode+and+habitat+use.%22+&btnG=.

14. bit.LY/TJOS-17-9; https://researchonline.jcu.edu.au/6807/1/6807_Choat_et_al_2003.pdf.

Chapter 18

1. bit.LY/TJOS-18-1; https://www.history.navy.mil/research/library/research-guides/z-files/zb-files/zb-files-p/perry-matthew-c.html.

2. bit.LY/TJOS-18-2; http://www.robinsonlibrary.com/america/unitedstates/naval/perry.htm.

3. bit.LY/TJOS-18-3; https://www.history.navy.mil/our-collections/photography/numerical-list-of-images/nhhc-series/nh-series/NH-65000/NH-65196.html.

4. bit.LY/TJOS-18-4; https://books.google.com/books?id=Tt4ZAAAAYAAJ&q=%22logbook%22+%22USS+north+carolina%22&dq=%22logbook%22+%22USS+north+carolina%22&hl=en&sa=X&ved=0ahUKEwiCl5KHnM7QAhWF4iYKHRrdDycQ6AEIODAB.

5. In the days before cameras on every phone, and before cheap photocopying, it was common practice to have a younger seaman write up copies of every letter that went out from the captain. The purpose of this was to track the incoming and outgoing correspondence—a bit like the email thread you'd have with the central office in your business emails.

6. bit.LY/TJOS-18-5; https://play.google.com/books/reader?id=fA6pCgAAQBAJ&printsec=frontcover&pg=GBS.PT337#v=onepage&q=delos&f=false.

7. bit.LY/TJOS-18-6; http://www.ibiblio.org/pha/USN/1826/NavReg1826.html.

8. Such as the announcement of the death of Thomas Jefferson: "The President of the United States with the deepest solicitude and sympathy directs that the death of Thomas Jefferson be announced to the Navy and Marine Corps and that they unite with their fellow Citizens in manifesting their sensibility at the bereavement sustained in the departure of this venerable Patriarch of the Revolution" (July 4, 1826, 41).

9. bit.LY/TJOS-18-7; https://finding-aids.lib.unc.edu/03190/.

10. Many thanks to Katie Odhner, who spent time in the archives at Chapel Hill, reading old ships logs.

11. bit.LY/TJOS-18-8; http://findingaids.loc.gov/db/search/xq/searchMfer02.xq?_id=loc.mss.eadmss.ms003026&_faSection=overview&_faSubsection=did&_dmdid=.

Chapter 19

1. For even more insights into what makes some researchers really effective, see Sam Wineburg, *Why Learn History (When It's Already on Your Phone)* (Chicago: University of Chicago Press, 2018).

2. Geoffrey C. Bowker and Susan Leigh Star, *Sorting Things Out: Classification and Its Consequences* (Cambridge, MA: MIT Press, 1999).

3. Sam Wineburg and Sarah McGrew, "Lateral Reading: Reading Less and Learning More When Evaluating Digital Information," Stanford History Education Group Working Paper No. 2017-A1, October 6, 2017; bit.LY/TJOS-19-1; http://dx.doi.org/10.2139/ssrn.3048994.

4. Surprisingly, yes they do! Can you do a little research to find out more about this topic?

5. For a great guide about fact-checking mistruths and misstatements that flow past us, see Daniel Levitin, *Weaponized Lies: How to Think Critically in the Post-Truth Era* (New York: Penguin Books, 2017).

Chapter 20

1. For a version of this question, see Sam Wineburg, *Why Learn History (When It's Already on Your Phone)* (Chicago: University of Chicago Press, 2018).

2. http://wolframalpha.com/.

3. Why did I choose these particular guys as an example? I was wondering if they were all contemporaries and could have all met over a hypothetical dinner. Once you see their timelines, it's obvious that they were. In this case, knowing how to do a simple query that creates a visualization really enhances your understanding of timelines and contemporary events.

4. https://www.bls.gov/.

5. https://www.google.com/publicdata/directory; https://registry.opendata.aws/.

6. There are many articles that point this out, but for perhaps the best one, see James R. Dwyer, "Public Response to an Academic Library Microcatalog," *Journal of Academic Librarianship* 5, no. 3 (1979): 132–141. In this piece, Dwyer shows that the probability of reading a book varies as the *cube* of the distance from the catalog. That is, you're more likely to access a text if it's close (or fast) to get to it. Something that's ten feet away is far more likely to be used than something that's a hundred feet away. The difference is dramatic.

7. See Eric Schurman and Jake Brutlag, "The User and Business Impact of Server Delays, Additional Bytes, and HTTP Chunking in Web Search" (paper presented at

the Velocity Web Performance and Operations Conference, San Jose, CA, June 22–24, 2009).

8. I'm not kidding; you can read all about it, but be prepared to read Renaissance Italian. See Frederico Grisone, *Ordini di cavalcare: et modi di conoscere le nature de'cavalli* (1565), bit.LY/TJOS-20-1; https://books.google.com/books?id=0vxUAAAAcAAJ&printsec=frontcover&dq=Federico+Grisone&hl=en&sa=X&ved=0ahUKEwiSgtbXvbLbAhUKnFkKHQb8AYoQ6AEILTAB#v=onepage&q=Federico%20Grisone&f=false.

9. bit.LY/TJOS-20-2; https://blog.verisign.com/domain-names/verisign-domain-name-industry-brief-internet-grows-to-330-6-million-domain-names-in-q1-2017/.

10. https://vr.google.com/earth/.

11. Knowing how to use the Internet Archive is increasingly becoming a basic research skill. But that's material for the next book. Luckily, as a skilled online researcher, you have the ability to search for the information you need to use it effectively. To get to the Internet Archive, see https://archive.org/.

12. One such bit of regulation is the "right to be forgotten" ruling, which allows people and institutions to require Google to remove specific articles from search results about them. This is an effort to allow bad (or out-of-date) content to be unfindable, but the possibility of misuse is huge.

13. Last search tip of the book: when you see a fragment of language that sounds a bit different than everything around it, such as this phrase "like tears in rain," it's usually worth doing a quick search to understand what it means. Don't let the meaning slip away in your reading.

Index